UNDERSTANDING HUMAN EVOLUTION

Exercises and Simulations
Diskette Guide and Workbook

RONALD K. WETHERINGTON
Southern Methodist University

WEST PUBLISHING COMPANY
St. Paul   New York   Los Angeles   San Francisco

WEST'S COMMITMENT TO THE ENVIRONMENT

In 1906, West Publishing Company began recycling materials left over from the production of books. This began a tradition of efficient and responsible use of resources. Today, up to 95% of our legal books and 70% of our college texts are printed on recycled, acid-free stock. West also recycles nearly 22 million pounds of scrap paper annually—the equivalent of 181,717 trees. Since the 1960s, West has devised ways to capture and recycle waste inks, solvents, oils, and vapors created in the printing process. We also recycle plastics of all kinds, wood, glass, corrugated cardboard, and batteries, and have eliminated the use of styrofoam book packaging. We at West are proud of the longevity and the scope of our commitment to our environment.

Production, Prepress, Printing and Binding by West Publishing Company.

Printed in the United States of America
99 98 97 96 95 94 93 92 8 7 6 5 4 3 2 1 0

ISBN 0–314–00908–6

CONTENTS

Preface

This guide and workbook are designed for any course in Physical Anthropology or related disciplines which covers human evolution. It provides exercises and simulations to make some of the fundamental concepts in evolutionary studies more easily understood, and to make the application of these concepts to the solution of evolutionary problems more accessible to the student. Hence, these exercises are suitable for both the introductory course and, in some of the more detailed options available, for the more advanced course as well.

The computer programs and their accompanying exercises in this workbook are designed as independent modules: they may be operated in any desired order, depending on the sequence of topics in the course in which they are used. They are organized in two sets, as follows:

The first set covers evolutionary genetics. These are programs which illustrate fundamental principles of Mendelian and population genetics, particularly as these apply to humans. The focus here is on genes and their alleles, how they produce heritable characteristics, and how these characteristics change from generation to generation.

The second set covers evolutionary results. These are programs which illustrate the emergence of humans and the fossil record which reflects our evolutionary past. The focus here is on the similarities and differences among species, both through time and across continents.

The programs in each set have accompanying exercises in this workbook which the student will complete as assigned in class. Whether a particular program and exercise will be assigned, and when, will depend on the instructor. The level of detail required of the student may also vary, depending on the nature of the course itself.

What is important in each program assigned, however, is that the student gain a foothold in understanding human evolution and its products. As a biological species we have been participants in a process that connects us with others; as a cultural species we have participated in a process that makes us unique. We are the only species interested in and capable of understanding these processes. These programs will, hopefully, spark that interest and contribute to that understanding.

The current version of this material results from several semesters of feedback from classes which used earlier versions, and from thoughtful suggestions from colleagues. I am grateful for these. I am particularly appreciative of the students in my Human Evolution course who critiqued each exercise and recommended changes, and to my Teaching Assistants who assisted in the preparation of much of this material.

Installation

The programs are supplied on one 5¼ inch (360K) diskette as well as on one 3½ inch (720K) diskette. Which diskette you will use depends on the diskette drives available on your computer.

Before using the programs you should make a backup copy of the diskette you will be using, and use this copy to run the program. The original diskette may then be put away in case something happens to the diskette you will be using. Depending on the configuration of your computer, different instructions apply in making this copy. The steps which follow should work:

To begin with:

1. Turn your computer on.
2. If your computer has a hard drive, you will shortly see the drive letter which will look something like this: **C:\>**.
3. If your computer has no hard drive, you will be asked to insert your DOS diskette (the **D**isk **O**perating **S**ystem diskette which came with your computer) in Drive **A** (or **B**) and press ENTER. After you have done this, you will see the appropriate drive letter, as above.

To make a backup copy:

1. Type **diskcopy a: a:** (or **diskcopy b: b:**) and press ENTER.
2. The computer will ask you to insert the "source" disk into drive A (or drive B). At this point place the original disk in drive A (or drive B) and press ENTER.
3. After a short time the computer will ask you to insert the "target" disk into drive A (or drive B). At this time place a new disk into drive A (or drive B) and press ENTER.
4. Continue to follow the on screen instructions.
5. When complete the computer will ask if you would like to copy another disk. Type "**N**".

To install on a hard drive:

1. At the C prompt (C:\>) type **MD EVOLVE** and press ENTER.
2. At the C prompt (C:\>) type **CD EVOLVE** and press ENTER.
3. Now the last line on your screen should read "C:\EVOLVE>. At this point place the original disk in drive A (or drive B).
4. If the disk is in drive A type **copy a:*.*** and press ENTER. If the original disk is in drive B type **copy b:*.*** and press ENTER.
5. When the computer has finished copying the program, it has been installed on the hard drive.

To run the programs from a floppy disk:

1. Place your backup disk in drive A (or drive B).
2. Type **A:** if the disk is in drive A (or type **B:** if the disk is in drive B) and press ENTER.
3. At the A prompt (or the B prompt) type **EVOLVE**.
4. Press "Return" at each screen until you see the Menu of Exercise Sets, then select the exercise you want to run.

To run the program from a hard disk

1. The last line on your screen should read C:\EVOLVE>. If it simply reads C:\>, then type **CD EVOLVE** and press ENTER.
2. At the C:\EVOLVE> statement type **EVOLVE** and press ENTER.
3. Press "Return" at each screen until you see the Menu of Exercise Sets, then select the exercise you want to run.

Note: Many computers with 5¼" drives have one which is "high density" (720K) and one which is normal density (360K). Your diskettes will operate on either drive, except for Exercise 13 (in Set 2): this exercise creates a separate file for you which will only work on normal density drives. Likewise, computers with 3½" drives frequently have a "high density" (1.44 Mb) and normal (720K), and the same provision applies here. If you don't know what the drive density is, don't worry about it for any of the other exercises.

Introduction

The programs on the diskette are designed to run on an IBM or compatible computer which has a VGA monitor.

For the most part, you will use the computer keyboard just as you would a typewriter keyboard. When you are asked to enter numerical data, you may use the number keys immediately above the letter keys, as with a typewriter, or you may use the special numerical keypad -- usually separated on the right side of the keyboard. If you use this numerical keypad, be sure the "**NumLock**" key has been pressed first.

You will often be asked to press the "Enter" or "Return" key: This is the same as the carriage return key on an electric typewriter. On some keyboards it is designated by the symbol "↵".

In some of the programs, you make selections of options by pressing the Function Keys. The are the keys labelled "F1", "F2", etc. and are either in two rows on the left of your keyboard or in a single row across the top. Do not mistake these for the numerical keys.

Getting Started

Please follow the instructions in the preceding Installation section.

You will select exercises from each Exercise Set from its own **MENU**. You access the Menu for Exercise Set 1 (Evolutionary Genetics), or Exercise Set 2 (Fossil Hominids), type from the **MAIN MENU** which you see when you type **EVOLVE** and page through the initial logo and copyright screens. Select an Exercise Set by typing its <u>number</u> and pressing "Return".

When you complete an exercise on the computer, you will be returned to the MENU for that Set, where you may choose another exercise in the Set, choose the alternate Exercise Set, or "Exit" from the diskette.

Every effort has been made to make these programs "user friendly", allowing easy access and easy exit. Occasionally, however, an alien gremlin may enter your system and cause it to "lock up" on you. If this happens, remove your diskette, and hold down the "Ctrl" and "Alt" keys while you press the "Del" key. This will "re-boot" your computer without the need to turn the power off and then on again. However, all previous entries will be lost and you will have to begin again. This will sometimes happen if you have a "mouse" attached to your computer. If this is the case, before following step 4, above, try typing "mouse off" <Return>. This will de-activate the mouse.

Using this Workbook

So much for the mechanics. Here's a brief overview of how to use this material. These programs are interactive in two ways: First, most of the programs ask you to enter information, either in answer to questions or to simulate certain processes. You enter this information through the keyboard. Second, each program instructs you on particular concepts which you apply by completing written exercises in this workbook.

To get the optimum benefit from a program, first read the introductory comments on the subject in this workbook. These comments 1) tell you the objective of the program, 2) outline the terms and/or concepts which the program assumes you already know (and the program you should complete first if you don't know them), and 3) provide you with a brief discussion of the terms and concepts you will apply in completing the program.

You then enter the program and follow its instructions. In each program, you will be given feedback: when your entries are correct, you will be told; when incorrect, you will learn why. In many cases, you can repeat sections of a program without starting from the beginning. In some cases, particularly in the simulations, you can save your entries or print them on your printer. You may be asked to print and turn in your work, and to turn in this workbook for review by your instructor.

For Exercises 1 through 5, multiple choice quizzes are provided in the Appendix. As you complete each of these Exercises, you may wish to check your understanding by using these. The correct answers, with explanations, are provided.

The programs are self-contained, but their order is by and large sequential: a given program frequently assumes understanding of principles reviewed in the program immediately preceding it. If your course text covers this material, use it as well. If the course has laboratory or discussion sections, let all of these be reinforcing mechanisms. Finally: don't be afraid to ask questions.

Exercise Set 1: Evolutionary Genetics

This Set contains exercises covering Mendelian genetics, human population genetics, and the evolutionary forces which act upon genetic systems and change them.

When you select this Exercise Set, here is the Menu you will see:

EVOLUTIONARY GENETICS

Although each of the following exercises is independent, the order follows a logical sequence, and concepts illustrated in earlier ones are assumed to be understood in later ones. After each exercise, you are returned to this Menu for further choices.

To make your selection, type in the number of your choice and press the 'Return' (or 'Enter') key. This key is sometimes marked ◄┘ .

M E N U

1 Basic Concepts in Genetics
2 Segregation and Independent Assortment
3 Allele Frequencies and the Hardy-Weinberg Model
4 Natural Selection
5 Gene Flow
6 DNA and Mutation
7 Gene Drift
10 Go To Fossil Hominids Menu

Press <ESC> to Exit

YOUR CHOICE?

In this workbook, terms which appear in **boldface** are important. You should understand the meaning of these and remember them.

Exercise 1
Basic Concepts in Genetics

Objective: This exercise reviews the basic concepts of genetics at the cellular level. You will learn the meaning of the terms "gene", "allele", and "locus", and the distinctions between the paired concepts "dominant-recessive", "genotype-phenotype", and "homozygous-heterozygous". You will also learn the distinctions between mitosis and meiosis.

Discussion: Genetic information is carried on the chromosomes in the nucleus of each cell. Except for our reproductive cells (gametes), all normal cell nuclei contain chromosomes in pairs. In our species, there are 23 pairs of chromosomes (46 total) in each nucleus. One of these pairs determines <u>sex</u>: these are the **sex chromosomes**. In the <u>male</u>, this pair consists of a chromosome labelled "X" and a shorter one labelled "Y". The <u>female</u> sex chromosomes consist of a pair of "X" chromosomes. The remaining 22 pairs of chromosomes are known as **autosomes**. Illustrated below are the 46 human chromosomes arranged in pairs according to their sizes and other characteristics. This arrangement is known as a **karyotype**.

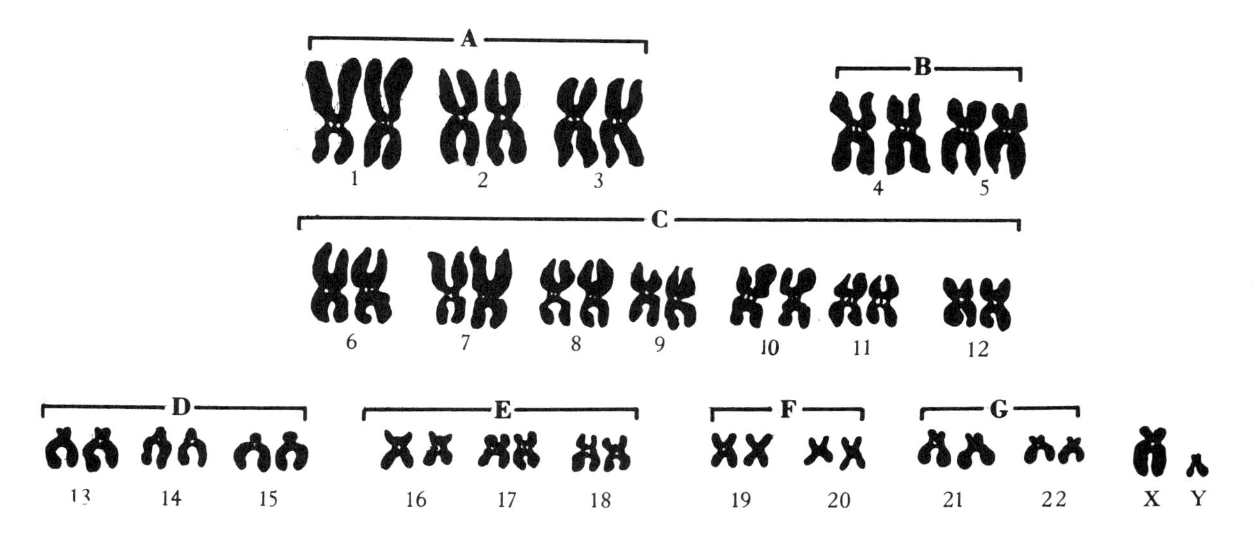

Since chromosomes are paired, so is the genetic information that is carried on them. The paired genetic information which is responsible for a particular characteristic is called a **gene**. Genetic information occurs at discrete points along the chromosome's length. These points are called gene locations, or **loci** (singular: **locus**). The individual paired loci, functioning independently as one locus or in groups of loci, determine the heritable characteristics of a species.

A characteristic which is determined by a single locus (with its paired information) is known as a **monogenic** (one-gene) characteristic, also called a **Mendelian** characteristic after Gregor Mendel, who made the initial discovery of the gene. For example, a person's A-B-O blood type results from paired information at the ABO-locus on a specific chromosome pair.

Each of the two genetic units of the pair contributes information to define the characteristic. These units are called **alleles**. An allele is a variant or alternative form that a genetic unit can display at a paired locus. An individual who is blood-type AB has one A-allele and one B-allele at the paired ABO-locus.

The two paired units at a locus together define the type of information present. This is known as the **genotype** of that characteristic. A person who has two A-alleles at this locus has an AA genotype; one who has an A-allele and an O-allele has an AO genotype.

If the two alleles at a locus are alike (e.g., two A alleles, or two B alleles, etc.) the genotype is said to be **homozygous** ("homo"="alike"). If the two alleles are different (e.g., one A and one O) the genotype is said to be **heterozygous** ("hetero"="unlike").

In some heterozygous genotypes, both alleles do not equally contribute to the characteristic. Rather, one allele may express its influence for the trait to the total exclusion of the other: in other words, it may dominate the paired locus. The allele which does this is known as a **dominant** allele, and the allele it dominates is called a **recessive** allele.

At the ABO-locus, allele O is recessive and both A and B alleles are dominant. In the AO heterozygote, only allele A functions and it prevents the expression of allele O. Since both A and B alleles are dominant, in the AB heterozygote they are co-dominant: each expresses itself.

In any heterozygous genotype where a recessive allele is present, the dominant allele determines the trait: a person whose genotype is AO is therefore blood-type A, just as if the person were homozygous AA. The expression of a trait is called its **phenotype**. Thus, both the homozygous AA and heterozygous AO genotypes are manifested, or expressed, as the same phenotype (blood-type A). A recessive allele only expresses itself when it is in a homozygous genotype (e.g., OO). Generally, a recessive allele is written as a lower-case letter and a dominant allele as a capital letter. At the ABO-locus, however, recessive O is capitalized.

The above discussion applies only to paired gene loci, of course: a paired locus on a chromosome pair will contain paired genetic information in the pair of alleles. All normal autosomes are paired in this fashion, and the linear points on one chromosome in the pair is matched by the same linear points on the other. The autosomes are **homologous** chromosomes. The sex chromosomes, however, are not. The male has one X and one Y chromosome. This is not a homologous pair -- the X is long, with many loci; the Y is short, with few loci. In the male, therefore, there are numerous single (non-paired) loci on the X chromosome. Alleles at these loci are alone and unpaired. Such a locus cannot produce either a homozygous or a heterozygous genotype. Its genotype, consisting of only a single allele, is said to be **hemizygous** ("hemi"="half"). Traits which are on this non-homologous part of the X chromosome are called **X-linked** characteristics, and they are expressed in the phenotype of the male even if the responsible allele is recessive. In humans, color blindness and hemophilia are examples of X-linked, or hemizygous, traits.

Be careful not to confuse X-linked loci with **linked loci**. The former refer to loci on the X-chromosome; the latter refer to two or more distinct loci that occur on (are linked to) any particular chromosome. After all, any chromosome has numerous loci; in humans there are at least 50,000 separate loci, on the autosomes as well as the sex chromosomes. Furthermore, it is apparent that genetic information on the sex chromosomes includes many characteristics which are independent of gender-determination.

Assignment: Run PART 1 of Exercise 1 on the diskette. This reviews the foregoing concepts and tests your understanding of them.

Now, let's move on: we have a lot of Mendelian characteristics whose phenotypes are determined by paired alleles on the autosomes, where alternative expressions are influenced by dominant and recessive interactions. We also have a few Mendelian characteristics whose phenotypes are determined by single alleles in hemizygous genotypes on X-chromosomes in males, and where dominance plays no role.

How does all of this genetic information get passed on from one cell to another in the living individual, and from one person to the other across generations? In both cases, the passing of information occurs through cell division, but of two different types.

Consider a person who is heterozygous AB at the ABO locus. Every body cell nucleus contains this paired locus on its paired chromosome. As the cell divides, tissues grow or are replaced. As a result of cell division, all new cells receive the same chromosome pairs, and thus this same heterozygous genotype. This form of cell division is called **mitosis**.

During mitosis, *each chromosome in the pair* replicates (copies) itself. In this example, the chromosome carrying the A allele replicates to form two, each with the A allele. These two identical chromosomes are called **chromatids**. The two chromatids separate and become incorporated into a different cell as the cell divides into two new cells with two new nuclei. The same process separates the replicated chromatids in the *other* chromosome of the pair: the chromosome carrying the B allele becomes two B-carrying chromatids, each incorporated into a different cell.

Hence, each final cell contains a chromatid carrying the A and a chromatid carrying the B. These are now called chromosomes once more, and represent a replicated pair. (Technically, two chromatids become chromosomes at their moment of separation from each other).

This mitotic cell division is graphically illustrated in the second part of the program on Exercise 1. While there are numerous recognized stages in this mitotic process, the general features are what are important. The graphic portrayal simplifies the process to point out these features.

While only one chromosome pair is illustrated, note that the final cells in the division all contain the full **complement** of chromosomes in pairs. This full complement is called **diploid** ("di"=2, "ploid"=chromosomes; denoting, therefore, a pair). This is also called the **2n** complement.

Note also that in the initial replicative stage of mitosis, when each chromosome in each pair has replicated itself, the *two* paired chromosomes are temporarily *four* chromatids. For this brief transitional time, therefore, the cell has a **tetraploid** or **4n** complement.

So far, this person with type AB blood has, through mitosis, produced new cells with the same genetic information as the original cell: this is how new cells are produced, how tissues grow, and how the body replenishes itself.

This person also produces **gametes**, the mature reproductive cells represented as sperm or egg cells. Gamete production, however, does not occur through mitosis. If a sperm cell carried the full (2n) complement of paired alleles, and it fertilized an egg which was also diploid, the resulting **zygote** (fertilized egg) would have a 4n complement, its offspring a generation later would receive an 8n complement, followed by a 16n complement, and so on, doubling each generation. Obviously, such a progression is not tolerable!

Instead, gamete production occurs through a cell division called **meiosis**. In meiosis, the complement is halved in the gametes, with each sperm or egg cell containing only one-half of each chromosome pair (and thus only one of the two paired alleles). Meiosis is for this reason also called **reduction division**.

In meiosis there are two cell divisions, and it is in the first of these that the reduction of chromosomes (and alleles) occurs. In this first division, the chromosomes align themselves by pairs rather than individually, for example, the chromosome carrying the A and the chromosome carrying the B are side-by-side as a pair. Replication occurs in each, but the chromatids remain united for a time.

As cell division occurs, one chromosome of each pair (e.g., the one carrying the A allele) is drawn to one side, the other chromosome (e.g., the one carrying the B) is drawn to the other, and when the new cells separate each cell has one-half of the original pair of chromosomes. The second division results when the replicated chromatids in each of these two cells separate, creating additional cells, which are the gametes. The gametes, therefore, unlike diploid cells, are **haploid**: they have a **1n** complement of chromosomes.

Hence, from an original cell with an AB heterozygote we find now four gametes: two each with a single A allele (on its single chromosome) and two each with a single B allele. The same holds true for each of the 46 chromosomes: each member of the 23 pairs separates from the other member into new cells, finally resulting in gametes with 23 single chromosomes each. Fertilization unites 23 from the sperm with 23 from the egg, and thus the zygote -- a new person -- has 46 chromosomes in 23 recombined pairs.

This meiotic division is graphically represented in the program for Exercise 1. Study both mitotic and meiotic cell divisions. Below are illustrated mitotic and meiotic divisions representing specific alleles. You should now be able to fill in these, similar to those in the program. Recognize that here we are dealing with autosomes only. X-linked loci would operate differently in meiotic division since these alleles are not paired.

By the way, is the karyotype illustrated earlier that of a male or a female?________

Assignment: Run PART 2 of Exercise 1 on the diskette. The following individual is genotype Tt at the Taster locus. Write in the alleles in each of the mitotic and meiotic stages:

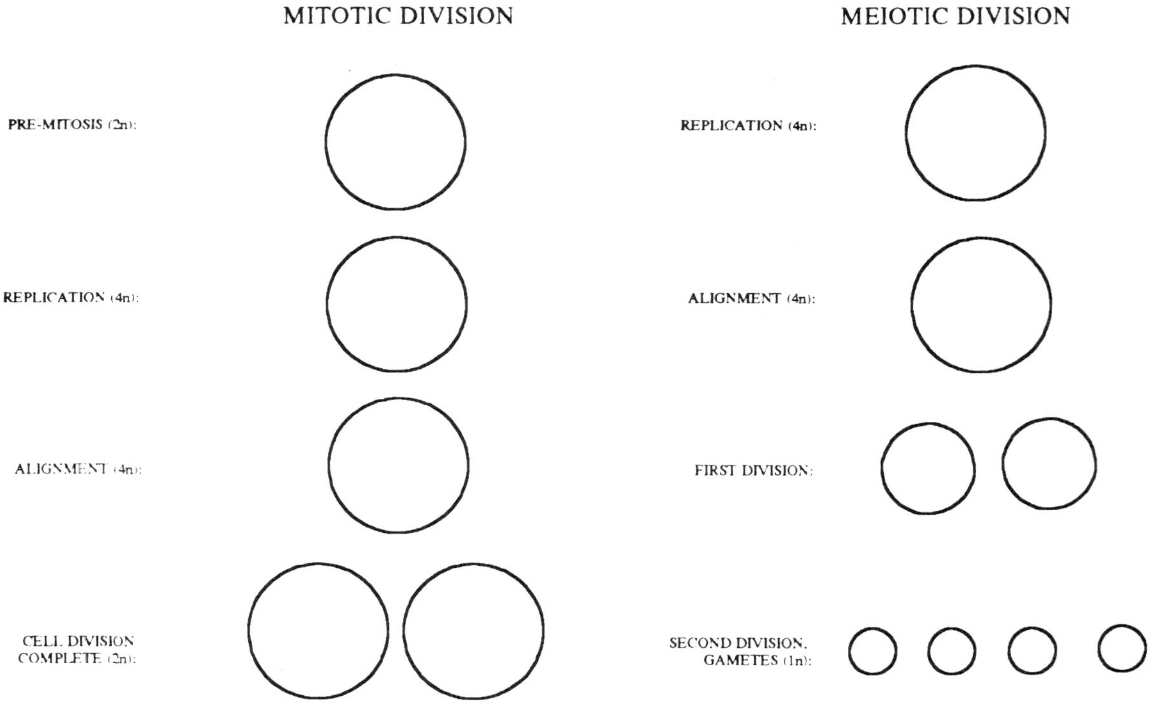

MITOTIC DIVISION

PRE-MITOSIS (2n):

REPLICATION (4n):

ALIGNMENT (4n):

CELL DIVISION COMPLETE (2n):

MEIOTIC DIVISION

REPLICATION (4n):

ALIGNMENT (4n):

FIRST DIVISION:

SECOND DIVISION, GAMETES (1n):

Exercise 2
Segregation and Independent Assortment

Objective: In this exercise you will apply the process of meiosis to examples of your own choice in order to describe and illustrate Mendel's fundamental principles of Segregation and Independent Assortment. This exercise assumes that you already understand the concepts covered in Exercise 1 and that you are familiar with the process of meiotic division in the production of gametes.

Discussion: Gregor Mendel, in his experiments in the 19th Century, demonstrated **particulate inheritance** -- that heritable characteristics are transmitted as discrete particles whose contributions from male and female unite during fertilization.

As you already know, an individual who is heterozygous at a locus (e.g., AB at the ABO-locus) will, through meiosis, produce gametes with only one of the alleles (e.g., some with allele A and others with allele B). Mendel recognized this separation of heritable units as critical to subsequent variation in the offspring. This is the **Principle of Segregation**: the two alleles at a locus will segregate into separate cells during meiosis.

Suppose the same individual is also heterozygous at another locus on another chromosome pair (e.g., Rr at the Rh-locus). At this locus, likewise, the alleles will segregate during meiosis: some gametes receive allele R and others allele r.*

This individual will thus produce 2 different kinds of gametes representing the ABO-locus and 2 different kinds of gametes representing the Rh-locus. With respect to *both* loci, this person will produce 4 different kinds of gametes: 1) **A** with **R**, 2) **A** with **r**, 3) **B** with **R**, and 4) **B** with **r**. The chromosome pairs, as they align pair-wise during meiosis, do so independently of one another.

Thus, the assortment of alleles at one locus into the gametes is independent of the assortment of alleles at the other locus into the same gametes. This observation Mendel also recognized as important in explaining variation among offspring. This is the **Principle of Independent Assortment**.

This means that during one cell's meiotic division, the alignment may result in the A-carrying chromosome being on the same "side" of the nucleus as the R-carrying chromosome (with the B and the r on the other "side"); but in another cell's meiotic division the A is on the same "side" as the r-carrying chromosome (with the B and the R on the other "side"). How the two pairs of chromosomes line up is mutually independent and random. Therefore, one alignment will occur as often as any other alignment as different cells undergo meiosis, and each of the different kinds of gametes will be produced with equal frequency.

---

*In our examples, the alleles **R** = Rh positive (Rh+) and **r** = Rh negative (Rh-).

Segregation and independent assortment can thus produce a great variety of gametes. The critical factor is whether loci are heterozygous. A homozygous genotype (e.g., AA) will segregate, but only 1 kind of gamete will be produced for this locus: all gametes will carry the A allele. Likewise, if two independent loci are each homozygous (e.g., AA and rr) they will assort independently but will not produce different gametes: all gametes will receive an A and an r allele.

In this exercise, you will create genotypes of your own choosing for a male ("Father") and a female ("Mother"). Your creation will then be subjected to meiosis and whichever Mendelian principles apply will be described for you. You will also be told how many different kinds of gametes each person will produce. You may elect to assign only one locus or you may assign two independent (unlinked) loci. You may also elect to assign genotypes only to Father or only to Mother. For Father's loci, you may utilize X-linked alleles and thus examine **hemizygous** genotypes.

If you include Mother in your choices, you will be quizzed over her gamete production after her meiosis is illustrated.

Assignment: Assign the following genotypes for Father and Mother and answer the questions in the program regarding Mother's genotypes:

1st Locus	2nd Locus
1. AA	Rr
2. AB	rr
3. AO	Rr
4. Tt	BO
5. TT	OO
6. AB	Tt

For each of the above cases, designate the phenotype and the number of different gametes the father can produce:

1st Locus	2nd Locus	# gametes
1.______	______	_____
2.______	______	_____
3.______	______	_____
4.______	______	_____
5.______	______	_____
6.______	______	_____

Exercise 3
Allele Frequencies and the Hardy-Weinberg Model

Objective: This exercise introduces population genetics. Here you will learn the concept of allele (gene) frequencies, and how frequencies at a single 2-allele locus may be determined for a population. You will then learn the Hardy-Weinberg model, which is the basic equation for examining the evolution of populations. This exercise assumes that you already understand the concepts and terminology which are covered in Exercises 1 and 2.

Part 1

Discussion: Evolution occurs when genetic variations -- including those produced by new allele combinations in fertilization by haploid gametes -- change through time. Such changes occur not in individuals but across individuals: in the transmission of alleles from parents to offspring. The biological unit which evolves is therefore not the individual but the population of individuals. When we study evolution we study populations -- as small as the family unit or as large as the species. The population is the unit of evolution; the individual is the unit of transmission of evolutionary change.

Genetic variations change through time whenever alleles appear, disappear, or occur in different proportions in a population from one generation to the next. This occurs whenever an individual reproduces -- or dies without having reproduced. The focus, however, is on alleles, not on individuals. Keep this focus in mind as we proceed.

In a population of 100 persons, each paired locus (2 alleles each) represents 200 alleles. At one locus, the proportion or percentage of all alleles which a particular kind of allele represents is known as its allele (or gene) **frequency**. For example, if at the ABO-locus everyone in the population is heterozygous AB, the A allele represents one-half of the 200 alleles: its allele frequency is .5 or 50%. If everyone is type-O (homozygous OO), the frequency of allele O is 1.0 or 100%.*

In order to determine whether evolutionary change is occurring or has occurred at a particular locus (or for a particular trait), we must be able to determine the allele frequencies at that locus. How we do this is quite simple.

Suppose we find, through blood typing, that this population has genotypes which are distributed as follows:

Genotypes:	AA		AB		BB		
	16 people	+	48 people	+	36 people	=	100 people
Genotype frequencies:	16%		48%		36%	=	100%
or:	.16		.48		.36	=	1.0

*Note that the meaning of "frequency" in population genetics is different from its more common usage in our language. In genetics, it refers to the proportion of the total represented by a particular unit.

What are the frequencies of alleles A and B? We can simply count them and divide (*Method 1*):

Genotypes:	AA	AB	BB		
	16 people +	48 people +	36 people	=	100 people
Allele A:	(2 x 16)=32 +	48		=	80 alleles
Allele B:		48 +	(2 x 36)=72	=	120 alleles
			Total	=	200 alleles

Thus the frequency of allele A is (80/200) or .40, and the frequency of allele B is (20/200) or .60 . Since each person has a pair of alleles, counting them requires division by the total to get the frequencies.

A simpler method is to get the proportions directly (*Method 2*):

An allele is represented in its homozygous individuals by 100%, and in the heterozygous individuals by 50%. Adding these proportions gives the correct frequency for each allele. For allele A, therefore, 100% of its homozygotes plus 50% of the heterozygotes = the frequency of A in the population:

Genotypes:	AA		AB		BB		
	.16		.48		.36	=	1.00
Allele A:	.16	+	.24			=	.40
Allele B:			.24	+	.36	=	.60

Assignment: Run Part 1 of Exercise 3 on your diskette and complete the problem you will find there. Then calculate the allele frequencies, using the same method, for the following populations:

Population 1: RR = 25 people Rr = 50 people rr = 25 people

Allele R = ______ Allele r = ______

Population 2: BB = .04 BO = .60 OO = .36

Allele B = ______ Allele O = ______

Population 3: TT = 160 people Tt = 480 people tt = 360 people

Allele T = ______ Allele t = ______

Part 2

Discussion: Consider once more the first population example we used.

Genotypes:	AA		AB		BB		
	.16		.48		.36	=	1.00
Allele A:	.16	+	.24			=	.40
Allele B:			.24	+	.36	=	.60

When this population reproduces, its next generation of offspring will result from the union of individual gametes, sperm and egg. Recall that our focus is on alleles, not individuals. Forty percent of all alleles at this locus in this population are A alleles; sixty percent are B-alleles. Their current residences are in persons -- paired in genotypes and segregated in gametes.

But these are temporary residences: the individuals are vessels which carry them for awhile before pouring them out. We can visualize the reproduction of this population as removing alleles from their individual vessels and placing them into a single reservoir -- a **gene pool**.

From this reservoir, alleles will be taken out and recombined, two-at-a-time (one from the male, one from the female) into new residences. This mating (taking alleles out in gametes and recombining them in fertilization) will be **random** for this ABO-locus: A new residence (baby) with a homozygous AA genotype will result when a A-carrying sperm fertilizes an A-carrying egg. How often will this occur?

It will occur as often as the gametes carrying these alleles will occur. Since the frequency of the A allele is .40 in this gene pool, 40% of all gametes will carry the A allele. The AA baby receives one A allele from the father and one A allele from the mother. Thus, 40% of all fertilizations will involve a sperm carrying the A, 40% of all fertilizations will involve an egg carrying the A, and 40% x 40% = 16% of all fertilizations will involve an A allele in both sperm and egg. The next generation of new residences will have a .16 homozygous AA genotype frequency.

The same principle applies to the combination resulting in the BB homozygote: With a B frequency of .60 in the gene pool, fertilization with a B-carrying sperm will occur 60% of the time, fertilization with a B-carrying egg will occur 60% of the time, and fertilization with both will occur 60% x 60% = 36% of the time.

For heterozygous AB recombinations, there are two opportunities: 1) a sperm with A (40% of all sperm) fertilizes an egg with B (60% of all eggs); and 2) a sperm with B (60% of all sperm) fertilizes an egg with A (40% of all eggs). The first opportunity represents a 40% x 60% = 24% frequency, the second opportunity represents a 60% x 40% = 24% frequency, and the two together represent a 24% + 24% frequency: heterozygous AB babies will represent 48% of the next generation's genotypes.

It is important that you understand the distinction between random mating when applied to *alleles* and when applied to *persons*: Persons seldom mate randomly -- we choose mates on basis of a wide variety of potential criteria. Most of our alleles, however, in their genotypes and manifested phenotypes, are not among these criteria. We are not, for example, more or less likely to choose a mate on the basis of that person's ABO blood type; nor is that person's ABO blood type more or less likely to influence his/her availability to be chosen. These alleles just happen to be in residence and they tag along. We mate randomly *with respect to* alleles at this locus.

This is not the case with all gene loci, of course. Until medical science resolved the problem, mating with respect to the Rh locus was <u>not</u> random, due to the danger of an Rh negative (rr) mother producing an infant with haemolytic anemia if the father was Rh positive (RR or Rr). More about this in Exercise 4.

Assignment: Run Part 2 of Exercise 3. Then fill in the frequencies for the population shown below.

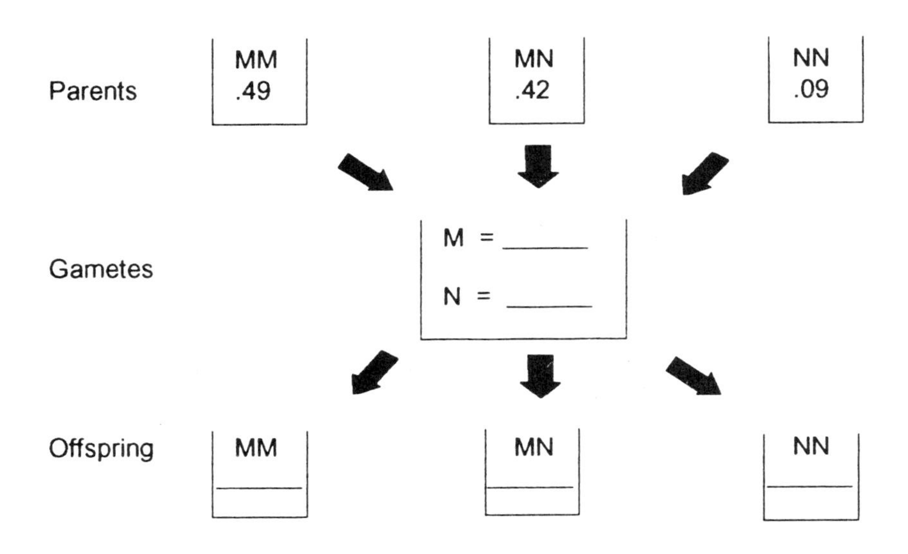

Discussion: You will notice that the distribution of genotypes among the offspring of this population is the same as that of the parental generation: the genotype frequencies do not change with random mating. Another way to examine this is with the Punnett Square, which is illustrated on the diskette.

Assignment: Following the illustration in Part 2 of the Exercise, fill in the Punnett Square on the following page with the allele and genotype frequencies for the population above.

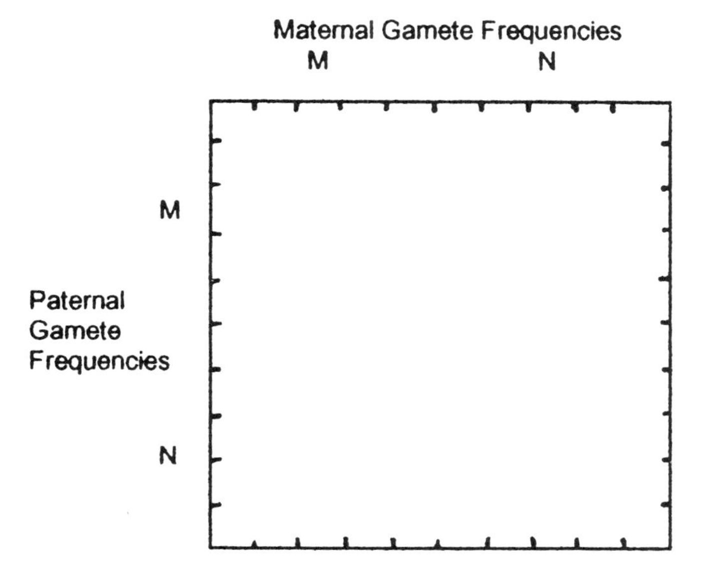

Discussion: Let's review these frequency relationships: When a population reproduces, any locus for which the mating is random will produce new genotypes in the next generation at frequencies which reflect the frequencies of the individual alleles. At a locus with two alleles in the population (e.g., A and B):

1) The frequency of each homozygous genotype will be the product of the frequency of its allele among the gametes (e.g., ^{f}AA = ^{f}A x ^{f}A and ^{f}BB = ^{f}B x ^{f}B).

2) The frequency of the heterozygous genotype will be twice the product of the frequency of each allele among the gametes (e.g., ^{f}AB = 2 x [^{f}A x ^{f}B]), because there are two ways a heterozygote may be produced in fertilization.

It is convenient to generalize this relationship for any random mating involving any two-allele locus: let p represent one allele and let q represent the other allele. If $p = .4$, as in our earlier example, and $q = .6$, (and thus $p + q = 1.0$), we find that:

A x **A** = ^{f}**AA** **B** x **B** = ^{f}**BB** 2(^{f}**A** x ^{f}**B**) = ^{f}**AB**

$p \times p = p^2$, and $q \times q = q^2$, and $2(p \times q) = 2pq$

$.4 \times .4 = .16$ $.6 \times .6 = .36$ $2(.4 \times .6) = .48$

Notice in this population that not only does $p = p^2 + pq$ (as in our Method 2 of counting allele proportions), but also $p = p^2$; and the same holds for q.

In other words, the relationship between allele frequencies and genotype frequencies in a random mating population is the simple algebraic relationship represented by the familiar binomial:

$$(p + q)^2 = (p + q)(p + q) = p^2 + 2pq + q^2 = 1.0$$

In evolutionary studies, this is known as the Hardy-Weinberg equation, and populations whose frequencies show this binomial relationship are called Hardy-Weinberg populations. This is named after the two scientists who first applied this equation to population studies in 1908, and it forms the basis for evolutionary genetics. (For loci with three alleles, the relationship is the trinomial, e.g., $(p+q+r)^2$, but the principle is the same).

This relationship will occur at loci for which there is random mating, although random mating <u>alone</u> does not necessarily mean that evolution is not occuring. When frequencies at a locus <u>do</u> show change, evolution at the locus is occurring, and the allele-to-genotype relationship is not in binomial equilibrium.

Therefore, one may evaluate the evolutionary status of a 2-allele locus in a population by applying the Hardy-Weinberg equation to it: if the locus is in equilibrium (not evolving), the square root of one homozygote + the square root of the other will = 1.0; if the locus is undergoing change (or is the result of change from the previous generation), this sum will not equal 1. The allele frequencies in this case may be calculated only by Method 2: $p = p^2 + pq$ (the homozygous genotype frequency plus one-half the heterozygous genotype frequency).

Assignment: For each of the following loci, calculate the allele frequencies from the genotype frequencies and mark whether the population is in Hardy-Weinberg equilibrium:

	Genotype frequencies	Allele frequencies	H-W? (X)
1.	AA = .04, AO = .32, OO = .64	A=______, O=______	______
2.	MM = .25, MN = .26, NN = .49	M=______, N=______	______
3.	RR = 49%, Rr = 42%, rr = 09%	R=______, r=______	______
4.	TT = .81, Tt = .18, tt = .01	T=______, t=______	______
5.	DD = .01, Dd = .10, dd = .89	D=______, d=______	______
6.	AA = .64, Aa = .20, aa = .16	A=______, a=______	______

The following populations are in Hardy-Weinberg equilibrium at the autosomal loci indicated. What are their genotype frequencies?

Known frequencies	Genotype frequencies
1. Rh negative phenotype = 4%	RR=_____, Rr=_____, rr=_____
2. Allele M at the MN locus = .2	MM=_____, MN=_____, NN=_____
3. Rh positive phenotype = .75	RR=_____, Rr=_____, rr=_____

Exercise 4
Natural Selection

Objective: In this exercise you will learn the meaning of natural selection as an evolutionary force and how it operates. The concept of fitness will be illustrated, and you will learn how to measure it in a population. The two channels through which selection operates -- fertility and mortality -- will be explained. Finally, you will run simulations of selection to illustrate the several ways selection influences phenotype frequencies. You should have a fundamental understanding of the concept and calculation of allele frequencies and the nature of the Hardy-Weinberg equilibrium, covered in Exercise 3, before doing this exercise.

This exercise is in two parts: Part 1 covers the nature and measurement of natural selection in populations. Part 2 covers the several kinds of selection commonly recognized.

Part 1

Discussion: Natural selection is the process by which phenotypes which are more adaptive, or more fit to survive in an environment, become more frequent in a population, while phenotypes which are less adaptive become less frequent. The result is generally an increase in the frequency of alleles which contribute to more adaptive phenotypes and a decrease in alleles which contribute to less adaptive phenotypes.

Such changes can occur only from one generation to another (not within a generation) as a result of the reproductive process. Changes in phenotype frequencies resulting from selection occur either through differences in fertility (adaptive phenotypes may promote higher fertility, while non-adaptive phenotypes may promote lower fertility) or through differences in mortality (adaptive phenotypes may have a lower rate of mortality, while non-adaptive phenotypes may have a higher rate of mortality). Hence, selection occurs either through **differential fertility** or through **differential mortality**, or both.

Note that the emphasis in selection is on <u>phenotypes</u>, not genotypes or alleles. This is because selection is in reality a measure of the interaction of a genetic characteristic and its environment. If the interaction is favorable to survival of the characteristic in the environment, the characteristic will remain or become more common; if not, it will be reduced in frequency or eliminated.

For a characteristic to interact in this way, it must be manifested. It is the phenotype which is manifested (apparent, or visible, or capable of reacting). The result of selective influence on the phenotype is a change in the frequency of the genotype(s) and allele(s) which are responsible for the characteristic. While selection operates on the manifestation of these, it is analyzed in reference to genotypes.

Selection is measured as the relative probability or frequency of survival among genotypes. This measure is either designated as relative <u>negative</u> survival (selection against a trait) or relative positive survival (fitness of a trait).

Fitness is therefore a term used as a mirror image of selection. For example, if at a locus one phenotype has a 25% lower survival rate than any other phenotype at that locus, the genotype(s) responsible are said to have a **selection coefficient** of .25 (a 25% disadvantage relative to any other genotype). It may also be said that the genotype(s) responsible for this 25% lower survival rate have a **fitness coefficient** of .75 (a survival rate which is 75% of the survival rate of the most adaptive genotype(s)).

Let's illustrate this. Suppose that the dominant/recessive allele frequencies at the A-a (albino) a population provide genotypes with the following frequencies:

AA(p^2)	Aa(2pq)	aa(q^2)
.36	.48	.16

This population is in Hardy-Weinberg equilibrium: it is not evolving. The genotypes were produced in these frequencies through random mating. (**A**=.6 and **a**=.4)

Let us assume that a fatal disease is suddenly introduced, to which the recessive phenotype is 25% more susceptible than is the dominant phenotype. Selection thus eliminates the recessive phenotype (and its genotype) at a rate which reduces the survival of this genotype by .25 each generation. Here is what we find in the first generation after the disease is present:

	AA(p^2)		Aa(2pq)		aa(q^2)		
Before Selection:	.36	+	.48	+	.16	=	1.0
Selection:	36 - 0 ---		48 - 0 ---		16 - 25% ---		
	36	+	48	+	12	=	96
After Selection:	(36/96) .375	+	(48/96) .50	+	(12/96) .125	=	1.0

If, instead of selection coefficients for these genotypes (0, 0, .25) we used fitness coefficients (1.0, 1.0, .75), the results would be the same:

Fitness:	36 x 1 ---		48 x 1 ---		16 x 75% ---		
	36	+	48	+	12	=	96

Here, selection has reduced the frequency of the recessive genotype. By definition, the other two genotypes have increased in their frequencies. Selection has also reduced the frequency of the recessive allele: from (.24 + .16)=**.4** to (.25 + .125)=**.375**, in this single generation. The dominant allele obviously increases as the recessive allele **decreases**. This is not because of any special advantage which the dominant allele confers, but rather because of **the** special <u>dis</u>advantage which the recessive allele confers through its **vulnerable** phenotype: it is vulnerable to the disease.

Below is a graphic plot of the recessive allele over 15 generations in this population, assuming that the selective disadvantage of the recessive phenotype remains at .25 (that is, assuming that the disease continues and this phenotype is affected by it).[*]

```
        Fitness values: AA=  1 Aa=  1 aa=  .75

                                       a
        0-------------------------------------------------100   Freq. a
1       >  .                 *                               .3750
2       >  .               *                                 .3522
3       >  .              *                                  .3315
4       > .              *                                   .3126
5       > .             *                                    .2954
6       > .            *                                     .2797
7       > .           *                                      .2653
8       > .           *                                      .2522
9       > .          *                                       .2401
10      > .          *                                       .2290
11      > .         *                                        .2187
12      >.         *                                         .2093
13      >.         *                                         .2005
14      >.         *                                         .1924
15      >.        *                                          .1849
        0-------------------------------------------------50
                                 Delta p
        * = p     . = Delta p
```

NOTE: In the following assignments, you run plots such as this for 10 generations. If you print these plots on your printer, run them for 20 generations to allow you to see the changes more dramatically. On your screen, running more than 10 generations will often scroll the top of the plot off of your screen.

---

*It is conventional, at the **p-q** locus, to represent the recessive (for illustration) as the **q** allele. In our simulation program, (but NOT in the **above example**), the plot showing frequency change across generations is **always** the **p** allele. This allows **a** more consistent basis for **comparison** of one population with another. Whether the **p** allele is dominant or recessive in a simulation, of course, depends upon the **fitness** values assigned to the three genotypes. The simulation thus allows sufficient flexibility to be useful in **illustrating** traits with *partial* **and** *incomplete* dominance in courses which reach this level of detail.

Note also that the program plots two sets of values: 1) the value representing the *frequency* of the **p allele each generation**, and 2) the value representing the *rate of change* in allele **p** (or **delta-p**) each generation. Examination of this rate of change **helps you visualize** the relative intensity of selection each generation. Your instructor may ask you to evaluate this separately from the change in the frequency of **p** .

Part 2

Discussion: The example we have just seen, selection against a recessive phenotype, is only one of several ways by which selection can change allele frequencies at a Mendelian gene locus. Selection may also operate against a dominant phenotype; against a heterozygote when it is phenotypically distinct from either homozygote; and against both homozygotes.

Examples of each of these alternatives may be seen in humans, and each affects allele frequencies in unique ways. We will examine these, in turn, as you run simulations on the computer which illustrate them.*

1. Selection against a recessive phenotype. Let's begin here, since the preceding example applies. You will note that the reduction of the recessive allele in that example becomes slower with each generation. In fact, the recessive allele will never be completely eliminated by selection, regardless of how lethal the condition causing the selection. This is because the recessive allele is always protected from removal when in the heterozygote.

Assignment: Run the simulation program for each of the following populations. Follow the instructions on your screen carefully. Note that in all of these simulations you will use fitness values instead of selection values: a fitness of 0 means that selection is 100% against a genotype, while a fitness of 1.0 means that selection is 0.

Generations: 10
Frequency *p*: .9 [the recessive allele]

	Fitness PP	Fitness PQ	Fitness QQ
1)	0	1	1
2)	.1	1	1
3)	.25	1	1
4)	.5	1	1

At the end of 10 generations, compare the frequencies of allele **p** in each population.

a) In which of the 4 populations do you find the greatest reduction in allele **p** ?

b) In which do you find the least change? Why?

c) Would you expect greater or less change in **p** if the fitness of **PP** were .75? Why?

d) The first population has the lowest fitness of the recessive homozygote **PP**: there is 100% selection against it. Does it also have the lowest allele frequency after 10 generations? If you continued reproducing this population for 20 or more generations, would allele **p** be eliminated? Try it, and explain the results.

*In this simulation program, you may plot the output (as shown on the previous page), or generate either a **graph** or a **table** of the frequencies. If you have a printer, you may also print some of these results.

In the human species, there are numerous deleterious recessive alleles such as this. These include albinism, cystic fibrosis, phenylketonuria, and Tay-Sachs disease. While these and other recessive conditions recur through random mutations, their frequencies in human populations are in large part due to the protection of the recessive allele in the heterozygote, which is phenotypically normal.

2. Selection against a dominant phenotype. Now we reverse the dominant/recessive relationship: allele **p** is now dominant rather than recessive. A dominant allele is expressed as a phenotype both in its homozygous genotype and in the heterozygote. If the heterozygote contains a recessive allele with the dominant, as in these examples, both genotypes are essentially identical as phenotypes: if selection operates against the dominant characteristic, it operates equally against both of these genotypes. This means that the dominant allele is systematically eliminated because it is always exposed when it is present.

Assignment: Run the simulation for each of the following populations.

Generations: 10
Frequency *p*: .9 [the dominant allele]

	Fitness PP	Fitness PQ	Fitness QQ
1)	0	0	1
2)	.1	.1	1
3)	.25	.25	1
4)	.8	.8	1

a) What happens to allele **p** in the first generation in each population? Why is there such a dramatic change in population 1 ?

b) How many generations does it take to eliminate allele **p** in each of these populations? [Try populations 3) and 4) with more than 10 generations each].

Deleterious dominant conditions in the human species include Chondrodystrophic dwarfism, congenital cataracts, hyperuricemia (gout), retinoblastoma (eye cancer), and Huntington's disease.

Of course, numerous normal conditions are also dominant. These include Rh-positive blood, blood types A and B, curly hair, and freckles.

3. Selection against the heterozygote only. This represents a rather unique kind of selection: It obviously requires the heterozygote to be phenotypically distinct from both homozygous genotypes. The most logical assumption is therefore that the heterozygote in such selection contains two **co-dominant** alleles.

For example, at the ABO-locus, there may occur three different heterozygotes: **AB**, **AO**, and **BO**. Only the first contains two dominants, and it is distinct from the AA andBB homozygotes. These others are not phenotypically distinct: AA and AO comprise a single phenotype, as do BB and BO. Hence, for selection to single-out the heterozygote in this case, it would have to be the **AB** heterozygote, right?

The answer is: not necessarily. Certainly, selection can operate against a co-dominant heterozygote, and such phenotypic distinction is clear-cut.

When selection operates against the heterozygote but not against the homozygotes, what happens? Here, both alleles are removed instead of just one. But if there are only two alleles at such a locus, **p** and **q**, both cannot be eliminated or even reduced in frequency without eliminating the population itself!

Assignment: Run the simulation on each of the 5 populations listed. Use the following fitness values for each simulation:

Generations: 10
Fitness, PP: 1.0
Fitness, PQ: .5
Fitness, QQ: 1.0

Initial Frequency of p:

1) .45
2) .49
3) .50
4) .51
5) .55

a) Compare the frequency of **p** in Population 1 with the frequency of **q** (which is **1-p**) in Population 5, after 10 generations of selection. Do the same for the 2nd and 4th populations. Explain the results.

b) When you run the 3rd population, examine the table of allele and genotype frequencies (the program gives you this option). Why do the values remain in equilibrium despite selection against the heterozygote?

c) What would happen in these five populations if the fitness of the heterozygote were 0 or .9, respectively, instead of .5? Explain.

d) Examine the table of allele and genotype frequencies for a population other than the third one during 20 generations: What happens to the frequency of the heterozygote in relation to the two homozygotes?

Despite the logic in assuming that this kind of selection can occur only when the heterozygote is co-dominant, the best example of this in our species is not with co-dominant alleles: it is at the **R-r** locus where the Rh blood type is determined.

Consider the following case: A woman who is Rh negative (genotype **rr**) conceives a child with a man who is Rh positive (genotype **RR** or **Rr**). The zygote receives a dominant **R** allele from the father, and is therefore Rh positive (genotype **Rr**).

To the mother's immune system, the synthesis of the Rh+ antigen by the developing fetus represents an alien substance: as these antigens pass into her from the placenta, her system synthesizes antibodies to fight them. The immune reaction this generates destroys the fetal red blood cells, threatening the infant's life. The infant suffers from a hemolytic anemia called *erythroblastosis fetalis.* While this process generally takes too long to affect her first pregnancy, the build-up of antibodies will be strong enough to destroy the cells of subsequent Rh+ pregnancies.

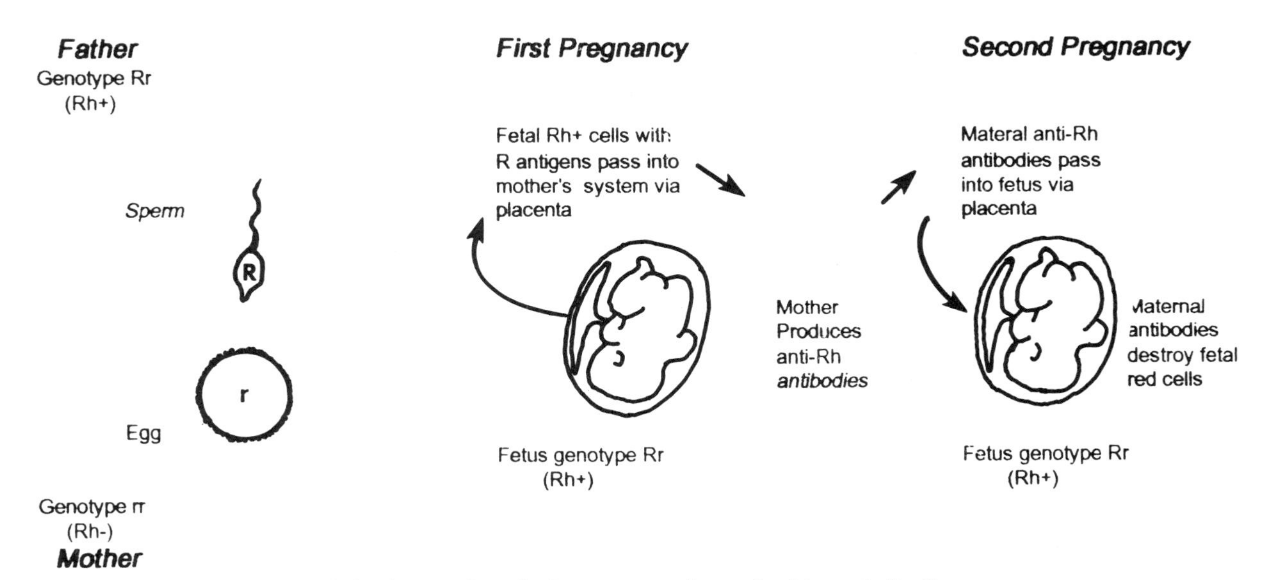

Selection against the heterozygote in erythroblastosis fetalis.

Selection has operated here against the <u>heterozygote</u>. The homozygous **rr** baby (Rh-) is not affected, since the mother is also **rr**; The homozygous **RR** genotype is non-existant among her pregnancies, since she will always contribute a recessive allele.

The disadvantage of the **Rr** heterozygote in this case <u>only</u> applies if the <u>mother</u> is genotype **rr**: In a case where the father is **rr** and the mother is Rh+, a heterozygous fetus (**Rr**) is not phenotypically distinguishable from the dominant homozygote (**RR**).

In this example, therefore, the phenotypic distinction between the heterozygous and the homozygous dominant genotype lies in the <u>environment</u> of the genotype, and not in the genotype alone.

This example demonstrates vividly that selection operates on the *phenotype* through the relationship between the *genotype* and the *environment.* Often, as in this case, whether a genotype is phenotypically distinct depends upon the environment in which it exists.

4. Selection against both homozygotes. Here is our final example of the ways selection may operate on a Mendelian locus. In many ways, it is the most important of them all.

In all of the previous examples, the net effect of selection is to significantly reduce (and in one case, eliminate) the frequency of one allele. If, however, selection operates on a locus by operating against both homozygotes, then by definition the most favorable (or adaptive) genotype is the heterozygote. This is a unique circumstance: the same alleles which are selectively removed when they occur alone (as genotypes **PP** or **QQ**), are favored when they occur together (as genotype **PQ**).

Furthermore, neither allele can be eliminated by selection, regardless of how deleterious their homozygotes may be. This is because their negative values as homozygotes must eventually come into an equilibrium, or a balance, with their positive value as a heterozygote. When this equilibrium is achieved, the allele frequencies do not change further.

For this reason, this form of selection is known as a **balanced polymorphism**. To get a better feeling for how this happens, it will be best to run our simulations first. We will run these for 50 generations each, so you may want to print both the plots and the tabular data on your printer.

Assignment: Run each of these four simulations. Read the following discussion after running the first simulation; then do the remainder and answer the questions:

Generations: 50
Fitness, PQ: 1

	Frequency p	Fitness PP	Fitness QQ
1)	.85	.75	0
2)	.4	.2	.2
3)	.3	.5	0
4)	.3	0	.5

The first example reflects the condition of sickle-cell anemia in some malarial regions of tropical Africa. Here, the allele responsible for the abnormal hemoglobin, **S,** causes sickle-cell anemia in the homozygote (**SS**), which is lethal in the absence of modern medical treatment. The normal homozygote (**AA**) is susceptible to malarial infection. The heterozygote (**AS**) has a high resistance to malaria and normally shows only mild symptoms of anemia.*

A fitness of 0 (selection 100% against) for the **SS** homozygote and a fitness of .75 (selection 25% against) for the **AA** homozygote creates the balanced polymorphism..The relative fitness of the heterozygote is thus 1.0 (selection is 0%).

*The S allele is considered an autosomal recessive by some. Here we treat it as codominant. The affects are the same.

The balance, or equilibrium, is achieved when selection changes the allele frequencies to **A** = .8 and **S** = .2; with these frequencies, continued selection will not alter the frequencies. Examine this illustration:

Allele frequencies:	A = .8	S = .2	
Genotype frequencies with random mating:	AA	AS	SS
	.64	.32	.04
Removal by selection (s_{AA}=.25, s_{SS}=1.0):	-16		-04
	-----	-----	-----
	48	32	0 = 80
Genotype frequencies after selection:	(48/80)	(32/80)	
	.60	.40	0

Allele frequencies: A = .6 + .2 = .8 S = 0 + .2 = .2

Note that once the initial frequencies of the alleles have achieved these frequencies, they remain constant despite continued selection, so long as the selection values themselves remain unchanged. The frequencies of the alleles are balanced by the selection values, which we may represent with the following diagram:

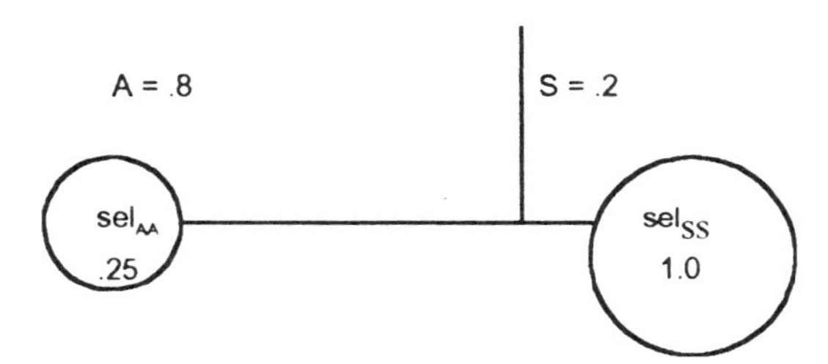

Note the relationships of the selection coefficients ("weights") and the allele frequencies ("arms") of this "mobile": The selection coefficients of .25 and 1.0 are in a ratio of .25:1 or 1:4 here. Their respective allele frequencies are in a ratio of .8:.2 or 4:1 . Here is the balance: change the frequency of either allele and it will return to this stabilization ratio given the selection ratio. Change the selection coefficient of either homozygote and the new ratio will cause a frequency shift until the alleles are re-stabilized under the new ratio. This is a balanced *polymorphism* because selection here guarantees that this locus will remain polymorphic: all three genotypes will continue to be produced each generation. The equilibrium equations for this relationship are simple:

$\mathbf{p} = s_3/(s_1 + s_3)$ and $\hat{\mathbf{q}} = s_1/(s_1 + s_3)$, where

p and **q** are the frequencies of **p** and **q** at equilibrium, s_1 is the selection coefficient of genotype $\mathbf{p}^2$ and s_3 is the selection coefficient for genotype $\mathbf{q}^2$.

Assignment: Now run the remaining simulations.

a) In your <u>first</u> simulation how many generations does it take to achieve the equilibrium of A=.8, S=.2? (Frequencies of .8003 and .1997, respectively, may be considered nominally in equilibrium).

b) What are the selection ratios of the other examples?

c) Populations 3) and 4) have opposite fitness values. What are the equilibrium (final) frequencies of **p** in each case? Does it take the same number of generations to reach these frequencies in each case? Explain.

Discussion: The operation of natural selection in populations, through differential fertility, mortality, or both, can occur in several different ways: selection may operate a) against a recessive phenotype, b) against a dominant phenotype, c) against a heterozygous phenotype only, and c) against both homozygous phenotypes. These classic examples, of course, illustrate selection at a Mendelian locus with two alleles. However, more complex traits with multiple alleles or multiple loci may also reflect these modes of selection.

In evolutionary studies, these forms of selection are frequently described in three categories: 1) **directional selection**, 2) **disruptive selection**, and 3) **stabilizing selection**. These categories correspond roughly to the modes we have discussed, and the simulations illustrate them very well.

Directional selection is selection which tends to make a variable trait less variable, or **monomorphic** (as opposed to polymorphic). It occurs when the most adaptive phenotypes are those at one extreme or another: at a dominant-recessive locus, this occurs when selection operates against the dominant phenotype or against the recessive phenotype. In the first case, the dominant allele is eventually eliminated, resulting in a totally monomorphic locus; in the second, the recessive allele is systematically reduced to the point where the dominant phenotype has become almost the only one. You can see these effects in the simulations. The same effects occur in complex traits under directional selection: assume that one extreme of stature has the advantage over all others. If the selective values are high, either very tall or very short people will become most common, shifting the range of stature in that direction. Darwin's explanation for the systematically longer-necked giraffe, through evolution, is an example of directional selection.

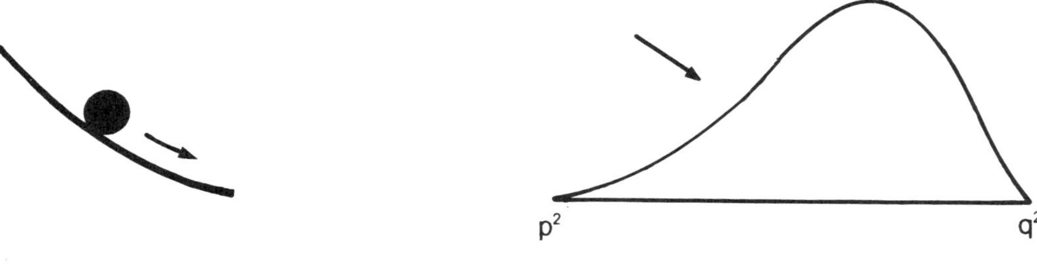

Schematic representations of directional selection

Disruptive selection occurs when separate adaptive advantages favor both ends of a continuum of variation (if, for example, both very short and very tall persons were optimum phenotypes, with the mid-range being selected against). In complex traits, this leads to polymorphism by guaranteeing variation; in a Mendelian trait, as we saw in the Rh example, it tends to reduce polymorphism by the disproportional removal of the allele with the lower frequency. Populations under selection which favors both homozygotes at the expense of the heterozygote may, in some cases, become diversified into two unique groups, or sub-populations, each adapting to their environment in different ways. This can lead, potentially, to mutual isolation and eventual species separation.

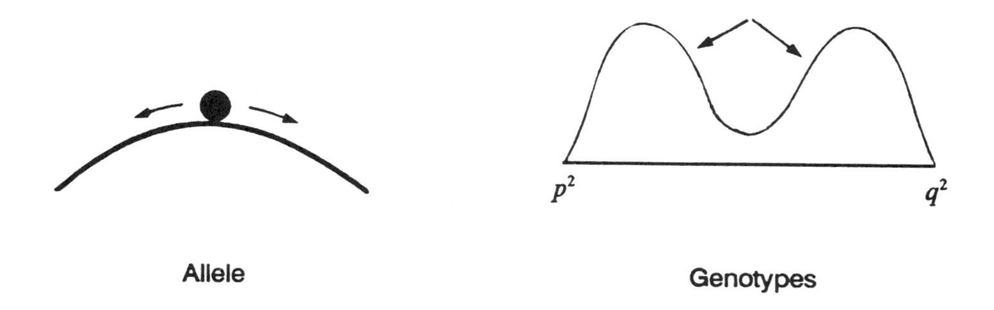

Schematic representations of disruptive selection

Stabilizing selection is the opposite of disruptive selection: it leads to polymorphism not by optimizing the morphological extremes, but by optimizing the middle road of variation. In our examples, the balanced polymorphism illustrates this. The extremes tend to be eliminated systematically as they are systematically produced by the optimal heterozygote. While the resulting gene pool shows frequencies which appear never to change, they are in fact in a dynamic equilibrium -- shifting each generation to achieve **homeostasis** by pushing the extremes out while maintaining the alleles responsible for the extremes. This typifies the process of evolution itself: evolution is a continuing process of matching of changing phenotypes to changing environmental demands. It is opportunistic: responding to conditions *as they are*, not as they may become. The reservoir of variations maintained through balanced polymorphisms provides ample alternatives for future change while providing ample opportunities for current survival.

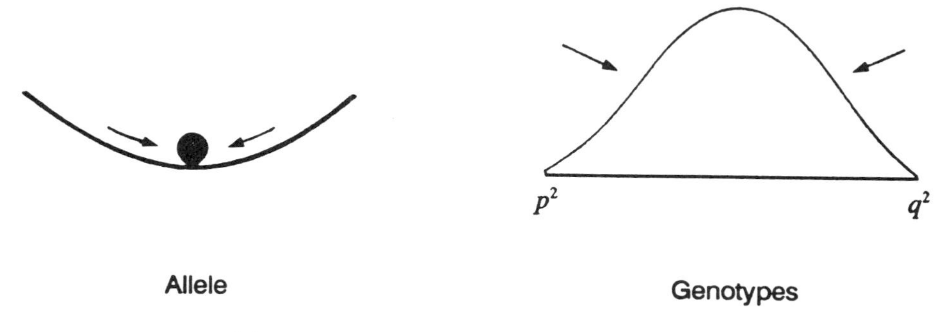

Schematic representations of stabilizing selection

Stabilizing selection may be seen in a number of <u>quantitative characteristics</u>. These are characteristics which are determined or influenced by alleles at several loci, and the resulting phenotypes display a quantitative, or continuous, variation. Stature is one example: stabilizing selection tends to reduce the relative frequencies of very tall and very short individuals, so that average stature in most populations is somewhere in the middle range.

Birth weight is perhaps the best example of stabilizing selection in humans. Extremely low birth weight babies have a higher mortality than normal-weight babies. Likewise, mortality rates are higher in high birth-weight babies. An optimum birth weight for most populations is between 7.5 and 8 pounds, and this is maintained by stabilizing selection against the extremes. The illustration below shows the relationship between birth weight and neonatal mortality (mortality of newborn infants under 28 days old).

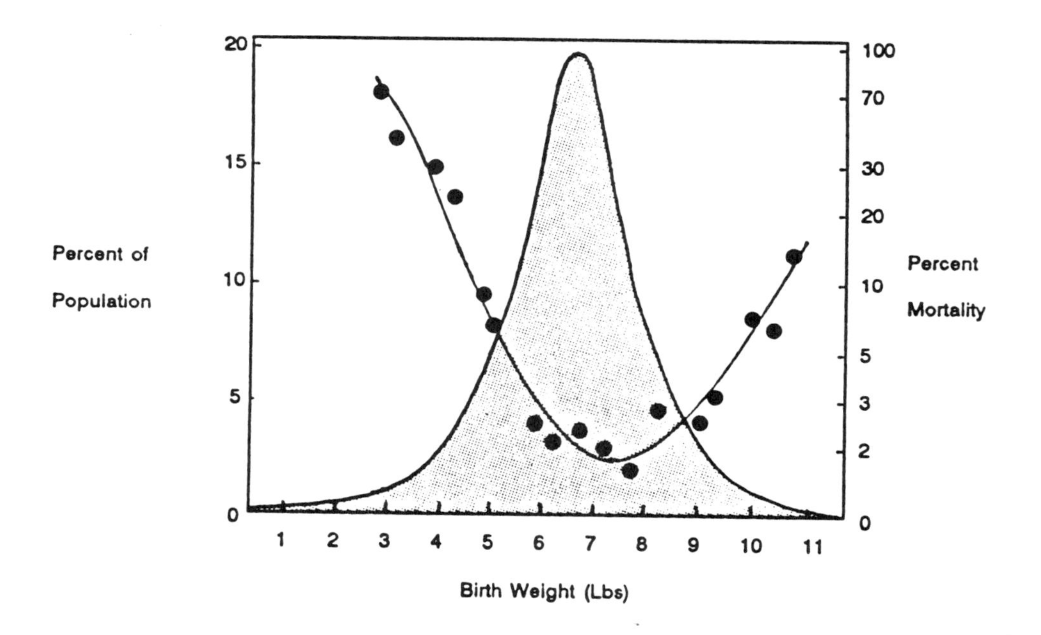

Birth weight and mortality

The distribution of birth weights by percentage of the population at each weight is represented by the shaded curve and the left-hand scale. The distribution of mortality at plotted birth weights is represented by the solid circles and the logarithmic right-hand scale of percent mortality. Here, the optimum birth weight is that with the lowest mortality. (From Cavalli-Sforza and Bodmer, 1971. Based on data from Karn and Penrose, 1951).

Exercise 5
Gene Flow

Objective: In this exercise you will learn the nature of gene flow as an evolutionary force, how it's effects are measured, and the several ways in which gene flow may occur in human populations. Before doing this exercise you should have an understanding of the concepts and calculations covered in Exercise 3.

Discussion: Gene flow occurs whenever alleles from outside a population are added to its gene pool. This most commonly occurs through marriage with outsiders. Because much of gene flow in human history has involved the movement of population segments into occupied territories, gene flow is also frequently called **gene migration**. In fact, most gene flow in human populations consists merely in systematic intermarriage among members of two or more populations, in which case a more appropriate term for gene flow would be **gene exchange**.

A third form of gene flow occurs when two or more populations (or representatives thereof) merge to form a new, distinct gene pool. The new population is a hybrid of its components, with gene frequencies reflecting their several contributions. This form of gene flow is called **hybridization**.

In all of these examples, the evolutionary process is the same: allele frequencies change by the introduction of new alleles from outside. While gene flow was probably always present in our species, it has played an increasingly important role in changing allele frequencies (and thus phenotypes) in our populations since the rise of civilization (ca. 5,000 B.C.).

Trade, conquest, the formation of alliances, and other cultural practices have encouraged interaction between adjacent groups. Such interaction inevitably results in interbreeding. As a result, gene pools continually change as people marry into and out of them.

On the other hand, religious, political, economic, linguistic and other cultural differences have established cultural barriers to gene flow even in the absence of natural barriers: Long-standing and deeply imbedded differences between the Islamic and Judaic populations of the Middle East have been effective in preventing gene flow between them, and thus in maintaining the mutual integrity of their gene pools.

Physical barriers such as the Great Wall of China and the Berlin Wall have from time to time been added to natural geographic barriers in halting the flow of people and their alleles across populations. When these walls are breached, the flow of alleles resumes.

Let's examine the action of gene flow, beginning with hybridization.

Assume that two populations, **I** and **II**, live some distance apart and have never interbred. Assume that one now migrates to join the other to form a single, hybrid population:

	I	+	II	=	H

Now let's assume that allele **R** at the Rh (**R-r**) locus has a frequency of 90% in population **I** and 20% in population **II**:

	I	+	II	=	H
R =	.9		.2		

The frequency of **R** in the hybrid population, since it now consists of the combined gene pools, will be a combination of the two contributing frequencies. What combination will it be?

It depends on the relative contribution that each population makes to the resulting hybrid. What if population **I** has, say, 200 people and population **II** has 300?

	I	+	II	=	H
N =	200	+	300	=	500
R =	.9		.2		

Of the 500 people in the hybrid group, population **I** contributes (200/500) = 40%. Likewise, population **II** contributes (300/500) = 60%. These are the relative contributions, called the **weighting factors** (**w**), of the alleles each population provides to the hybrid:

	I	+	II	=	H
N =	200	+	300	=	500
w =	.4	+	.6	=	1.0
R =	.9		.2		

The frequency of allele **R** in the hybrid will be the sum of the frequency of **R** in each contributing population times its proportionate contribution of **R** (its weighting factor):

	I	+	II	=	H
N =	200	+	300	=	500
w =	.4	+	.6	=	1.0
R =	(.9 x .4)	+	(.2 x .6)		
=	.36	+	.12	=	.48

If the two contributing populations were equal in size (regardless of the actual number of people), each would contribute equal proportions to the hybrid population, and the weighting factor for each would be 50% .

We can now represent gene flow as a standard formula with respect to allele **p**:

$$p_H = (p_I \times w_I) + (p_{II} \times w_{II})$$

Assignment: In 1789 a mutiny occurred aboard the H.M.S. Bounty as it sailed in the South Pacific. Mr. Fletcher Christian, the first mate, together with 24 other crew members, put the captain and the rest of the crew in a dinghy and cast them off. The British sailors, under Christian's command, sailed the Bounty to nearby Tahiti, where they resided for a short while.

Then Christian and 8 sailors, together with 12 Polynesian women, sailed to uninhabited Pitcairn Island. Here they scuttled the Bounty and founded their new population. Only 6 sailors lived to reproduce. The descendants of these 18 original inhabitants live on Pitcairn today, having had little migration in or out since 1790.

Can you estimate the composition of the original Pitcairn gene pool? Consider the following:

1) At the ABO-locus, the current frequency of allele **A** in England is 20%; in Tahiti its frequency is 40% .

2) Assume that these frequencies were the same in 1790.

3) Assume that these were the frequencies represented, respectively, by the original British and Polynesian settlers of Pitcairn.

Apply the gene flow equation to these data, labelling and filling in the blanks below:

British mutineers		Polynesian women		Pitcairn gene pool	
(_____ x _____)	+	(_____ x _____)	=	_____	(freq. A)

Similar examples of relatively insulated hybrid gene pools have occurred throughout human history, although most such hybrids are not geographic isolates. In colonial America, the slave trade brought in black Africans from the Gold Coast. The consequent interbreeding with white plantation owners created, in the African-American, a hybrid gene pool reflecting this gene flow.

Similar circumstances have defined Ladino gene pools (European white and native Indian admixtures) throughout Latin America. In South Africa, where Bantu-Dutch interbreeding created such a hybrid population, Apartheid legally recognized, isolated, and defined this gene pool: the Cape Coloured.

Most human populations exposed to gene flow do not form hybrid gene pools which can be identified and isolated. Instead, populations regularly and routinely exchange alleles through varying patterns of intermarriage from generation to generation. Consider again our first example:

Populations	**I**	**II**
Allele R =	.9	.2

Assume that these two populations do not merge into a single hybrid. Instead, each generation, population **II** receives some alleles from population **I** through intermarriage, and likewise population **I** receives some from population **II**. In other words, the populations remain separate but gene flow occurs between them.

Let's assume that population **II** has a 10% intermarriage rate with Population **I**. That is, 10% of all marriages in population **II** involve a mate from population **I**. This means that as population **II** (the "recipient" population in this case) reproduces, 10% of its marriages are with individuals migrating in from population **I** (the "donor" population).

The *gene flow rate* (the flow of alleles) is ONE-HALF this intermarriage rate: every such marriage combines ONE person in population **II** with ONE person from population **I**. Thus the gene flow rate is 5%: Within population **II**, 5% of its gene pool are alleles "flowing" in from population **I** and the remaining 95% of its gene pool comprises its resident alleles. These proportions are the weighting factors for the next generation's frequencies:

Population II_1		**Population I**		**Population II_2**
$(R_{II} \times w_{II})$	+	$(R_I \times w_I)$	=	R_{II}
$(.2 \times .95)$ +		$(.9 \times .05)$		
.19	+	.045	=	.235

In summary, here is what this represents:

1) Population **II**, with a frequency at the Rh locus of **R** = .2, produces its next generation mostly from within: .95 of its offspring alleles come from its own gene pool.

2) However, 10% of its marriages each generation come from population **I**, with one marriage partner migrating in.

3) The gene flow rate from population **I** is therefore 5%: .05 of the offspring alleles in population **II** come from population **I**, where the frequency is **R** = .9 .

4) As a result, the frequency of **R** in the offspring generation of population **II** is 23.5% (up from 20%). So in the next following generation, the initial **R** frequency is this frequency.

What will happen to the frequency of **R** in population **II** after more generations of gene flow? Obviously, if it began with a 20% frequency and continues to receive 5% of its alleles from a population with 90% **R**, its frequency will increase. For how long?

Assignment: Run the simulation program for population **II** through several generations. In the simulation let **p** = **R** (the preceding example), let the **donor** be population **I**, and the **recipient** be population **II.***

How high will the frequency go? Will its increase be at a constant rate or not?

Run each of the following gene flow examples using the values provided:

Frequency of **p** in recipient population: **.1**
Frequency of **p** in donor population: [see below]
Proportion of donor migrants (gene flow): [see below]
Generations: **16**

	Donor frequency p	Proportion donors
1)	1.0	.05
2)	1.0	.1
3)	1.0	.5
4)	0.2	.1
5)	0.2	.2
6)	0.2	.5

1. Compare the first 2 populations: the second has twice the gene flow rate of the first. Does it also have twice the frequency of **p** after 16 generations? Explain.

2. The third population has 50% gene flow -- every mating involves one member from the donor population. What is the maximum frequency which allele **p** can achieve? When does this occur? Why doesn't it occur in the first generation?

3. For populations 4 through 6, what is the maximum frequency allele **p** can achieve? In each case, how many generations does this take?

4. Population 5 has twice the flow rate of population 4, but the same donor frequency. Compare the differences in recipient frequency through the 16th generation. When does the greatest change occur in each, and at what point are the differences between the two the largest?

A final note: gene flow is seldom one-way. Normally, each population is both a donor and a recipient of alleles at a relatively constant rate per generation. For simplicity in both illustration and in operation, the simulations here illustrate one-way gene flow. Give some thought to what the effect would be in the simulations above if the rates of gene flow applied mutually to the two populations.

---

*In this program, you may continue any plot for *n* generations, beginning where the previous generation left off: thus, you may run one simulation for consecutive 10-generation periods rather than for a single 20-generation period. This helps prevent the screen from scrolling the data out of sight before you visualize it. You may also print the output on your printer.

Exercise 6
DNA and Mutation

Objectives: In this exercise you will 1)learn the nature of mutation at the molecular (DNA) level, and 2) learn how to measure its evolutionary influence at the population level.

The exercise has three parts which are mutually independent and which support these objectives:

PARTS 1-2 support the first objective. They cover molecular genetics, including DNA replication and protein synthesis. They assume that you understand the fundamental concepts covered in Exercise 1, and may be used in conjunction with that Exercise.

PART 3 supports the second objective. It covers dominant and recessive mutations, the concepts of genetic and mutation load, and the calculation of mutation rates in a population. It assumes that you understand the concepts and applications of evolutionary genetics as covered in Exercise 3.

Part 1

Discussion: At the molecular level, mutation may best be understood as an *error in code transcription.* An allele is a unit of genetic information represented in the form of a code. This genetic code is defined in and by a sequence of chemical components which comprise a molecule of **deoxyribonucleic acid** (**DNA**). In more conventional terms, gene = DNA. It is more appropriate, however, to conceptualize a locus as a DNA molecule, and a paired locus as a pair of DNA molecules.

The DNA molecule replicates itself during mitosis/meiosis. It also transcribes (copies) its code to other molecules, **ribonucleic acid** (**RNA**). This transcribed code is used to synthesize proteins by causing **amino acids** (the basic components of proteins) to link, or bond, together in long chains.

A particular protein is defined by both the number and sequence of amino acids in its chain, from among the 20 amino acids available. If the number or the sequence of amino acids is altered, so is the resulting product. When this happens as the result of an error in DNA replication, or in DNA-RNA transcription, a **mutation** has occurred. Thus, *a mutation is a chemical alteration in protein synthesis.*

Now note this: DNA replication occurs both in *mitosis*, the body-building cell division which occurs in the great variety of tissues forming our bodies, and in *meiosis*, the gamete-creating cell division by which we transmit something of ourselves to our offspring. Mutation may occur at any locus during either of these cell divisions. Where it occurs, and when, become profoundly important to us individually and to our species.

Assignment: Run Part 1 of Exercise 6 while following the narrative below.

1) The DNA molecule as two major components: First is a double-helix (two entwined spirals) known as the **backbone**. Each backbone, consisting of phosphate-sugar bonds (the sugar is **deoxyribose**), is the same from one DNA molecule to the other.

2) The second component is the nitrogen **base** which is attached to the sugar of the backbone. The base, with its bonded sugar and phosphate, is called a **nucleotide**. The two backbones are joined together by their sequenced bases. These nucleotide pairs (called **base pairs**) are like rungs on a ladder (or better, since this is a double helix, like steps in a spiral staircase).

3) It is in the base pairs that DNA molecules differ from one another. There are only four kinds of bases, and thus four different nucleotides. Each nucleotide differs from the other three in the kind of nitrogen base it has.

4) Two of these, **Adenine** (**A**) and **Guanine** (**G**) are chemically similar and are known as **Purines**. The other two, **Thymine** (**T**) and **Cytosine** (**C**), are also similar and are known as **Pyrimidines**.

5) In forming these base-pair "steps" or "rungs", hydrogen bonds form the connection. The base pairs are quite specific: **Adenine** bonds only with **Thymine**, and **Guanine** bonds only with **Cytosine**.

6) When replicating (as during mitosis or meiosis), the hydrogen bonds break, the paired bases separate, and the backbones uncoil. As this happens, free nucleotides in the cell move in to bond with their complementary, and now exposed, bases on each unwinding backbone. Each backbone, then, serves as a **template** to define the sequence and number of free nucleotides as bonding progresses. The original DNA molecule has now formed two identical DNA molecules: each double-helix comprises one of the original backbones with its nucleotide bases and one newly-formed backbone.

Part 2

Discussion: The function of the gene (as a DNA molecule) is to synthesize proteins. It does so by transmitting its code to sites in the cell where proteins are formed. The genetic code consists only of four different units: the **nucleotide bases** (**Adenine**, **Guanine**, **Thymine** and **Cytosine**).

Proteins consist of linked sequences of **amino acids**. There are 20 of these which, in varying combinations and quantities, make up all of our proteins. It is obvious that each of these proteins must be uniquely coded by the four bases of DNA. The breaking of this code (by Frances Crick, 1961) came from some truly elegant experiments. It was as much a benchmark in revolutionizing biology as it was a hallmark in scientific method.

If the code contains only four letters, and if different combinations of these letters represent code words for the 20 amino acids, how many letters make up each code word?. Certainly more than one; and necessarily more than two -- since four letters taken two at a time = 4^2, or only 16 different words.

Three, however, are sufficient, allowing $4^3 = 64$ different code words. Experiments have verified this: the genetic code consists of a three-letter word, or a sequence of three nucleotide bases (called, appropriately, a **codon**). This code is illustrated on page 38.

Assignment: The transcription of the code from DNA to the final synthesis is portrayed on the diskette. Run Part 2 of Exercise 6 while following the narrative below.

1) During transcription, the double-helix unwinds -- as it does during replication. But in this case complementary nucleotides from a different source, **ribonucleic acid** (**RNA**), link together along the unwinding DNA template. This new single-strand of nucleotide bases *does not bond with the backbone template.*

2) Instead, it acts as a messenger, transmitting the transcribed code from DNA in the nucleus to small protein bodies in the cytoplasm called **ribosomes**. This form of RNA is appropriately called **messenger RNA** (**mRNA**).

3) RNA differs from DNA in two important respects (in addition to having a different sugar, ribose rather than deoxyribose): a) it is single-stranded, and b) it has **Uracil** (**U**) in place of Thymine. Uracil bonds with Adenine in RNA just as Thymine does in DNA.

4) This mRNA molecule, with its sequence of triplet codes, attaches to the ribosome as the latter "moves across" the mRNA strand.

5) Meanwhile, still another form of RNA "picks up" amino acids in the cytoplasm to transfer these to the ribosome. This **transfer RNA** (**tRNA**) consist of short, single strand molecules. Each bears a codon for a specific amino acid.

6) The tRNA molecules transfer their amino acids in sequence to the appropriate codon-translated sites on the ribosome. As soon as adjacent amino acids are bonded together in the protein-synthesis process, the tRNA strands are released. The protein has now been synthesized by the translation of the code.

At this level of analysis, mutations may be seen to occur in different ways. First, one amino acid may be substituted for another because of a transcription error. Second, in transcription, an mRNA strand may have either deleted or added a codon, thus changing the product synthesized. Some mutations produce codons whose presence during synthesis simply stops the process, while others act to supress codons which would stop the synthesis.

One of the most studied of genetic diseases is sickle-cell anemia. The A-S locus produces the SS homozygote with phenotypic distinctions at all levels of analysis, from the molecular to the whole person. At the genetic level, the S allele responsible for the abnormal protein results from a mutation of the normal hemoglobin-A allele (HbA). This mutation, it was discovered (by Ingram, 1957), results from a *single* amino acid substitution in one of the long chains of amino acids comprising hemoglobin:

HbA amino acid sequence:

Valine-Histidine-Leucine-Threonine-Proline-**Glutamic acid**-Glutamic acid- . . . .
1 2 3 4 5 **6** 7 146

HbS amino acid sequence:

Valine-Histidine-Leucine-Threonine-Proline-**Valine**-Glutamic acid- . . . .
1 2 3 4 5 **6** 7 146

You will note that in the sixth position in the sequence for HbS, Valine has been substituted for Glutamic Acid. There are two codons in DNA which code for Glutamic Acid: **GAA** and **GAG**. Of the four codons which code for Valine, two (**GTA** and **GTG**) involve a difference in only a single base: **Thymine** has replaced **Adenine** in the middle letter of the triplet code.

This difference, chemically, is insignificant. However, the effect of this single mutation on the several levels where it is phenotypically expressed becomes increasingly profound:

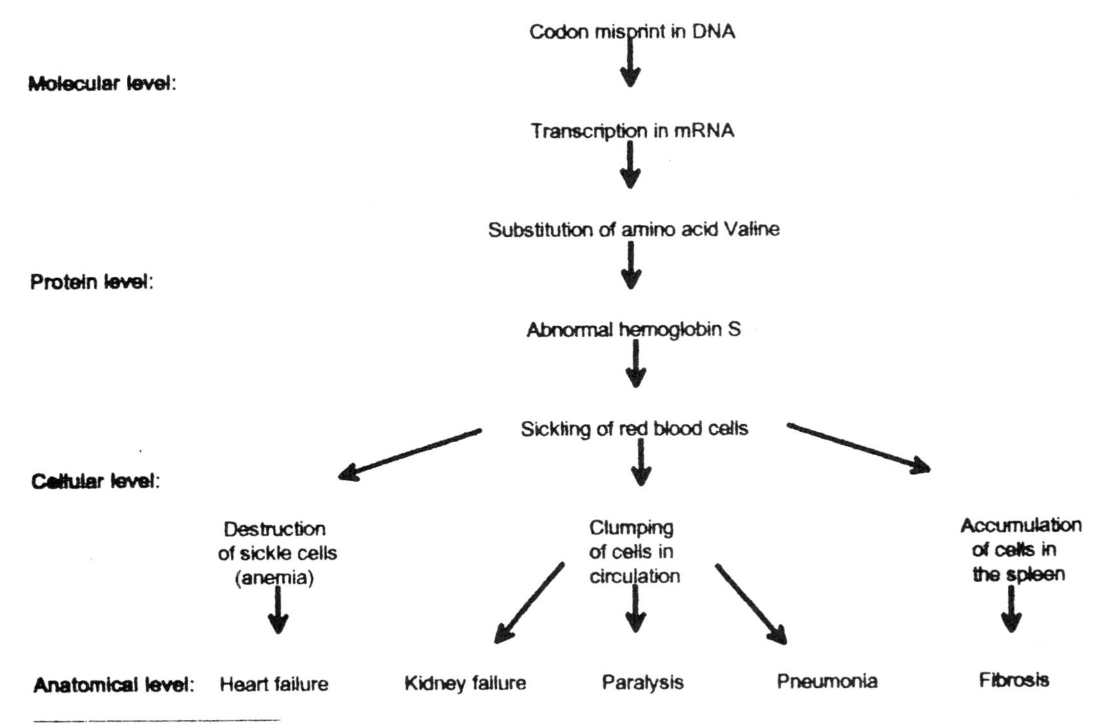

*See also Exercise 3, pp. 24-25.

Assignment: Solve the problem below, using the genetic code as shown and the appropriate base pair complementarities for DNA and RNA.

UUU, UUC: Phe UUA, UUG: Leu	UCU, UCC, UCA, UCG: Ser	UAU, UAC: Tyr UAA Stop UAG Stop	UGU, UGC: Cys UGA Stop UGG Trp
CUU, CUC, CUA, CUG: Leu	CCU, CCC, CCA, CCG: Pro	CAU, CAC: His CAA, CAG: Gln	CGU, CGC, CGA, CGG: Arg
AUU, AUC, AUA: Ile AUG Met	ACU, ACC, ACA, ACG: Thr	AAU, AAC: Asn AAA, AAG: Lys	AGU, AGC: Ser AGA, AGG: Arg
GUU, GUC, GUA, GUG: Val	GCU, GCC, GCA, GCG: Ala	GAU, GAC: Asp GAA, GAG: Glu	GGU, GGC, GGA, GGG: Gly

The Genetic Code[1]

Fill in the blanks, indicating the correct bases (transcribing and translating from left to right) and amino acids (note that the <u>bottom</u> DNA strand is the active one in the mRNA transcription):

	C			T				G	DNA double helix
T			A						
		G							mRNA transcription
			AAU						tRNA codons
						Arg			Amino acids coded

[1]These are the triplet codons for tRNA. Note the range of codons which identify particular amino acids (the acid names are abbreviated). Note also the three "Stop" codons: these terminate the synthesis at the ribosomes.

Part 3

Discussion: The probability that a particular mutation will occur within a particular time period is a measure of the **mutation rate**. This is most commonly estimated from the frequency of the mutant phenotype in a population: that is, from the proportion of the entire population which is represented by the mutants.

Consider, for example, the condition called **Chondrodystrophic dwarfism** in humans. This condition is a cessation of growth in the limbs and other bones at an early age. It results from a dominant mutation. The normal recessive **d** allele mutates to the abnormal dominant **D** allele.

A study in Denmark revealed that babies with the mutant phenotype are born at the rate of 10 in every 100,000 (10×10^{-5}) live births. On average, 2 of these 10 babies are born of parents of whom at least 1 is also a dwarf. Consequently, only the remaining 8 reflect new mutations.

The proportion 8/100,000, representing new mutants, is the proportion of mutant genotypes to normal genotypes at birth. The proportion of mutant to normal alleles is a direct estimation of the mutation rate at this locus:

$$\frac{8Dd}{100{,}000\ dd} = \text{GENOTYPES} \qquad \frac{8D}{200{,}008\ d} = \text{ALLELES}$$

The estimated mutation rate (from **d** to **D**) is thus approximately 8/200,000 mutant alleles at this locus, or 4/100,000, or 4×10^{-5}. Note that the 8 mutant genotypes are estimated as heterozygotes. This is because the likelihood of two identical mutations producing a homozygote mutant individual is extremely low.

Estimating recessive mutations is much more problematic and is indirect. Since a mutant recessive is undetectable in the heterozygote, a recessive phenotype resulting from a new mutation cannot be distinguished from a recessive phenotype resulting from the mating of two heterozygotes.

Assignment: The rate of mutation has been estimated for a number of deleterious human alleles, ranging from .5 per million ($.5 \times 10^{-6}$) to 5 per 100,000 (5×10^{-5}). Since the more common mutations are most likely to be detected, it seems likely that the actual average mutation rate for humans is at the lower end of this range, say one per million (10^{-6}).

Assume this to be the case. Taking the number of human (paired) loci at 50,000, the likelihood that an individual will produce gametes with one mutation is:

$$\frac{1}{1{,}000{,}000} \times 50{,}000 = \frac{50{,}000}{1{,}000{,}000} \text{ or } \frac{1}{20} = 05$$

Thus, the likelihood of a newborn receiving a spontaneous mutation from <u>either</u> the father <u>or</u> the mother is:

Father		Mother			
.05	x	.95	=	.0475	(from father, not from mother)
.95	x	.05	=	.0475	(from mother, not from father)
.05	x	.05	=	.0025	(from both father and mother)
			Total	.0975	

Thus, at this mutation rate, each of us has a 10% likelihood of carrying at least one mutation which occurred in and was inherited from one or both parents.

What if the mutation rate is actually 10^{-5}? Calculate:

Father		Mother			
____	x	____	=	_____	(from father, not from mother)
____	x	____	=	_____	(from mother, not from father)
____	x	____	=	_____	(from both father and mother)
			Total	_____	

As an evolutionary force, mutation is ultimately responsible for <u>all</u> allelic forms of a gene. These persist at higher frequencies when they are not deleterious: deleterious mutations will be eliminated (if dominant) or kept at very low frequencies (if recessive) through natural selection (as demonstrated in Exercise 5, Part 2).

While dominant mutations are visible in all genotypes containing them, recessive mutations lie hidden in the heterozygote and thus will accumulate through reproduction. Those that are deleterious, therefore, pose a potential threat to the gene pool. The sum of the threat of deleterious recessive alleles, then, involves both their reproduction and their spontaneous occurrence through mutations.

The effects of a particular recessive mutation on a population are very small if mutation alone (and not reproduction) is responsible. Even the accumulation of a harmless allele in a population, through constant mutation, is a very slow process.

Consider a population with a frequency of **A** (the normal allele) = 100%; consider further that the rate of mutation of this allele to the recessive **a** allele is 10^{-5}, or 1 in 100,000. With no selection against it, how long will it take the new allele **a** to achieve a frequency of, say, 10% (with **A** = 90%)?

A very long time! On the following page is a plot of frequency change in allele **A** given this mutation rate.

```
          PLOT OF ALLELE 'A' RESULTING FROM MUTATION TO 'a'
      BEGINNING FREQUENCY OF A =  1     MUTATION RATE A   a =   .00001
      FOR  13000  GENERATIONS (PLOTTING EVERY  1000  GENERATIONS)

Generation                                                   Frequency of A
         0--------------------------------------------------100
 0        >                                                  * %1.000
 1000     >                                                 *  .990
 2000     >                                                 *  .980
 3000     >                                                *   .970
 4000     >                                                *   .961
 5000     >                                               *    .951
 6000     >                                               *    .942
 7000     >                                              *     .932
 8000     >                                              *     .923
 9000     >                                             *      .914
 10000    >                                             *      .905
 11000    >                                            *       .896
 12000    >                                            *       .887
 13000    >                                           *        .878

  After  13000  generations, mutant allele "a" has a frequency of .122
```

You will note that Allele A reaches 90% (and mutant **a** 10%) only after 10,000 generations!

Assignment: Part 3 of this exercise on the diskette provides plots like the one above. Simulate mutation's effects on a population using the values given below. Since the change is so small each generation, a large number of generations must be simulated, and there is no need to plot the new frequency for every generation! Use the following values AND, when the program asks, plot every 1,000 generations:

Frequency of mutant allele a (from A) resulting from constant mutation rates:

BEGINNING FREQUENCY *A*	MUTATION RATE m (x 10^{-5})	NUMBER OF GENERATIONS	FREQUENCY OF *a* IN THE LAST GENERATION
1	5	15000	____________
.5	.1	20000	____________
.5	1	15000	____________
1	.1	100000	____________

The first example above represents the estimated mutation rate for albinism in one population. The second represents the rate for Huntington's disease (a dominant mutation). Note that a rate of .1 x 10^{-5} is the same as 1 x 10^{-6}, or one mutation in a million. Experiment with this simulation using other values.

Exercise 7
Genetic Drift

Objective: In this exercise you will learn the nature of genetic drift as a random evolutionary force, how it reduces variation in the population, and how its action is measured. Before you do this exercise you should understand how allele frequencies in a population are determined and the nature of Hardy-Weinberg equilibrium (Exercise 3).

Discussion: You will recall from Exercise 3 that in any population which, for a particular gene locus, is randomly mating will be expected to maintain its current allele frequencies in the following generation, so long as no evolutionary forces are influencing the locus. This is known as a Hardy-Weinberg population. The provision for such a population requires, therefore, that the population be reproductively isolated from others (there is no gene flow introducing into it or removing from it any alleles through interbreeding). It also requires that no mutation occurs at the locus, and that all of the phenotypes at the locus have equal advantage (no selection is occurring).

Given these provisions, the allele frequencies may still change in the next generation, due to the random nature of allele recombinations in forming zygotes. A couple who are both heterozygous at a locus may produce offspring who are all homozygous for one of the alleles; the other allele is thus not represented at all in their next generation. This change in allele frequencies is simply the result of the random sampling of alleles in the gametes of the parents, and is known as **genetic drift**.

Such change in a small population of two reproducing individuals is not at all unexpected or unlikely. If there were four individuals (two couples), all heterozygous, it would be less likely that their collective offspring would all be homozygous for one allele, of course, and if there were eight such an event would be even less likely. The measure of these likelihoods is a measurement of probability, and we can say that the probability of such dramatic change (from a 50% frequency of an allele to a 100% frequency) decreases as the population (sample size) increases.

Another way to say the same thing is to say that the magnitude of the change is less as the sample becomes large. Consider, for example, tossing coins: let each coin represent an individual who is heterozygous, possessing two alleles -- a head and a tail. Tossing each coin represents choosing the allele which the individual's gamete will contribute to the next generation. In the toss, each individual has an equal likelihood (50%-50%) of contributing each allele.

Assume that ten coins therefore represent ten individuals who will each "toss" a single gamete with its allele into the gene pool of the next generation. The equal likelihood of the two alternative alleles represents, therefore, a random sampling of the alleles among these ten heterozygous individuals. Since the sampling is random (the toss) and the alternatives are equal (either heads or tails), we expect that in the toss we willget five heads and five tails, thus continuing the 50%-50% frequency for the next generation. However, we would not be dismayed to find four heads (40%-60%) or four tails (60%-40%). In a single toss, even three of one and seven of the other would not be highly unlikely. Our sample, after all, is a very small one.

Now assume a population of 100 coins: tossing them likewise gives us an expectation of similar results from random sampling, but finding forty heads (40%-60%) or forty tails is certainly less likely, and finding thirty of one and seventy of the other is highly unlikely. We might intuitively assume, furthermore, that finding 45 of one and 55 of the other is probably about the same likelihood of finding 4 and 6 in our ten-coin sample (45%-55% vs. 40%-60%).

Note here that we are really comparing the amount of deviation from our expectation: not whether the frequencies will change, but rather how much they will change. Minor changes are more likely than major changes in both sampled populations, but if major changes do occur they are more likely to occur in the small population.

It is important to remember that gene drift is random: this means that the likelihood that the frequency of "heads" in the sampled gametes will decrease by, say, 10% is exactly equal to the likelihood that it will increase by the same percentage. Its actual decrease in a particular generation may therefore either increase in the following generation or decrease still further.

This is why this evolutionary force is called random drift. The frequency of an allele in a gene pool will drift up and down, back and forth, from generation to generation. The net effect of drift, however, is to reduce the amount of genetic variation in a gene pool. There are two ways to visualize this:

1) The maximum genetic variation in a population occurs when all of its alleles are equally represented. At a two-allele locus, frequencies of .5 for each allele define this maximum, which is represented in a Hardy-Weinberg population by a .5 frequency of the heterozygote genotype (and .25 frequencies for each of the homozygotes). The change in either allele frequency by drift necessarily reduces this variation.

2) An allele which exists at a low frequency in a population, for whatever reason, may be totally eliminated from the gene pool by random drift. For instance, an allele with a frequency of 1% has an equal likelihood of increasing to 2% or decreasing to 0%. When the latter happens, all variation has been lost -- not simply reduced! The smaller the frequency, the greater the likelihood that drift will eliminate it, regardless of population size.

Assignment: The program, Genetic Drift, on your diskette simulates this random loss of variation in a population. In it, you choose a) the population size, b) the initial frequency of **p** (at the **p-q** locus), and c) the number of generations you wish thispopulation to reproduce. The program then plots the frequency of **p** after subjecting it to the force of drift each generation. (Using random numbers, it selects either allele **p** or allele **q** from each individual's gametes to produce the next generation's offspring.)

Along with the graphic plot, the decimal frequency of **p**, and the amount of variation at this locus, are given for each generation.

Variation is expressed as the **standard deviation.** This is a statistical measure of the amount of variation in a characteristic. It decreases both as an allele frequency moves away from 50% and as the size of the population increases. It designates the magnitude of change in an allele frequency which will occur with predictability. Your particular course may or may not cover this concept in depth, and you may become familiar with the related concept of **variance**. Regardless, you may at least know this: as the frequency of **p** increases or decreases on either side of 50%, variation at the locus is reduced, reflected in a smaller value for the standard deviation. Ultimately, particularly in a small population, this value reaches zero (when one of the two alleles is by chance eliminated).

Use the following values in the program, and notice both the "wandering" fluctuation of allele **p** and its changing standard deviation:

Run #	Freq. p	Population Size	Number of Generations	Plot every *X* Generation(s)
1	.5	10	20	1
2	.5	20	20	1
3	.5	50	20	1
4	.5	100	50	5
5	.2	10	20	1
6	.2	20	20	1
7	.2	50	20	1
8	.2	100	50	5

Since this program simulates random events, no two runs with the same values will give you identical results. These runs (and others you may choose) illustrate the higher probability of an allele extinction if either a) the population is small or b) the initial allele frequency is small. Both a) and b) virtually guarantee extinction within a few generations. Over enough generations, however, even moderate-sized populations will eventually lose an allele.

Note that such evolutionary change in characteristics occurs here only as a result of random gamete sampling. Natural selection, mutation, and gene flow are not involved.

With the larger gene pools of the modern world, genetic drift obviously plays a much smaller role in evolutionary change. However, the major portion of our evolutionary past occurred in the context of small, relatively isolated human groups. We may infer that genetic drift played a more profound role in this early shaping of the human condition.

Printing your results: You may send your results to your printer in two ways. You may either directly print the plot you see on your screen, or you may print a new set of results using the same parameters (initial frequency, population size, number of generations, etc.) you just used for the screen display. The reason for these two methods is simple: Drift is a random event each generation, and no two simulations using the same parameters will be identical.

For example, suppose you selected a population size of 10, an initial frequency of **p** = .5, and let the program run through 20 generations, plotting each generation. Chances are fairly good that one allele (**p** or **q**) will become extinct before those 20 generations. Running these same parameters several times will result in some extinctions of **p**, some extinctions of **q**, and some with no extinctions. Furthermore, the inter-generational values will be different each time.

When you complete a simulation, at the bottom of the plot you will see the statement "To print these results, press the PrtScn (Print Screen) key". If you do this, the plot on your screen will be printed.

When you press **RETURN** (either after or instead of printing), you see the question "Do you want to run the program again, with NEW values, (Y or N)?". Answer Y if you want to do another simulation with a different population size, frequency, etc.

If you answer N, you see the question "Do you wish to send a new plot, using THESE values, to your printer (Y or N)?". If you answer Y, the program will run another simulation using the <u>same</u> population size, frequency, etc. which you just used. You may continue printing different simulations with these same values by repeating this procedure. The advantage to this is that it reveals how great or small the differences among several otherwise identical populations can be as the result of genetic drift. Multiple simulations of small populations, for example, will vary widely in allele frequency changes due to drift; multiple runs of large populations will vary much less.

When you answer N to this last question, you exit the program and are returned to the **MENU**. Be certain, if you print your results, that your printer is hooked up and turned on.

Exercise Set 2:
Fossil Hominids

This set contains exercises at the Macroevolutionary level. The first, Exercise 8, treats phylogenetic principles: species, subspecies, and their evolutionary relationships. The last, Exercise 13, invites you to reconstruct your own phylogeny. Between them, Exercises 9-12 provide graphic comparisons of fossils and their geographic distributions.

The graphic displays in this Set require the use of a VGA color graphics card and monitor. When you select this Exercise Set, here is the Menu you will see:

EXERCISES IN FOSSIL HOMINIDS

The following exercises cover Macroevolutionary concerns. They correspond to the same numbered exercises in the workbook, Understanding Human Evolution. After you complete an exercise, you are returned to this Menu for further choices.

To make your selection, type in the number of your choice and press the 'Return' (or 'Enter') key. This key is sometimes marked ◄┘

M E N U

8	Phylogenetic Principles
9	Comparative Hominid Crania
10	Hominid Maps: East Africa
11	Hominid Maps: Europe
12	Hominid Maps: Asia
13	Reconstructing Human Phylogeny
14	Go To Evolutionary Genetics Menu

Press <ESC> to Exit

YOUR CHOICE?

Exercise 8
Phylogenetic Principles

Objective: In this exercise you will learn the conventional principles used in classifying species and the problems involved in applying species and subspecies classification to fossils in reconstructing phylogenetic relationships.

Discussion: The exercises in Set I focused on how variation in human populations is influenced by the evolutionary forces of mutation, selection, gene flow, and genetic drift. We now begin an examination of the evolutionary background of our species, *Homo sapiens.* We are all members of this species, and it possesses a broad range of phenotypic variation. We see this variation within any one of our human breeding populations, even relatively isolated gene pools such as those of the Canadian Eskimo or the Kalahari Bushmen. Furthermore, differences in phenotypic variation distinguish these and other gene pools from one another as groups.

In other words, our species is both **polymorphic** (variation among individuals) and **polytypic** (variation between groups). Variation between groups allows identification of distinct regional or geographic characteristics conventionally categorized as "races". The term "race" has been misused in both biological and social contexts to distinguish groups possessing biological traits which are non-genetic and, inappropriately, as a term for ethnic differences. Characteristics which result from underlying genotypes are perhaps more appropriately defined as **subspecies** characteristics. Hence, a **subspecies** is a gene pool which differs from other gene pools of the species in particular genotypes.

When evolutionary forces result in a change in genotype frequencies, the variations which distinguish the particular gene pool shift -- some characteristics becoming less and others becoming more common. This evolutionary change is generally known as **microevolution**: changes in genetic variation within the species across generations, sometimes leading to identifiably new subspecies.

On the other hand, changes in genetic variation may become so substantial over successive generations as to lead to a new species. Evolutionary change of a magnitude which results in a new species (the process of **speciation**) is generally known as **macroevolution**. If we accept the assumption that the same evolutionary forces are responsible for both microevolutionary and macroevolutionary change, a problem in evolutionary classification becomes the determination of when evolutionary change has led to speciation.

For most biologists working with living populations, the species concept is relatively objective: two populations belong to the same species if they are capable of interbreeding and producing fertile offspring. Thus, **interfertility** becomes the primary criterion. If the two populations are **intersterile**, they belong to different species. If the two populations, through breeding isolation (lack of gene flow) have become identifiably different in their variations, but are nevertheless **interfertile**, they may be distinguished as subspecies but not as different species.

However, one cannot apply this species concept to extinct species: the potential interfertility of extinct populations cannot be measured. Instead, we must use more indirect evidence of their genotypic similarities and dissimilarities. Traditionally we have used skeletal morphology in distinguishing subspecies and species, and have used morphological similarities through time to reconstruct their evolutionary relationships.

Phylogeny, the evolutionary relationships among species, involves two kinds of evolutionary events: 1) the event of **succession**, when one species evolves into and is replaces by another in a single multi-generational lineage, and 2) the event of **adaptive radiation**, when one species subdivides or branches into several lines, each becoming separate species through mutual isolation and different adaptations.* These are illustrated below:

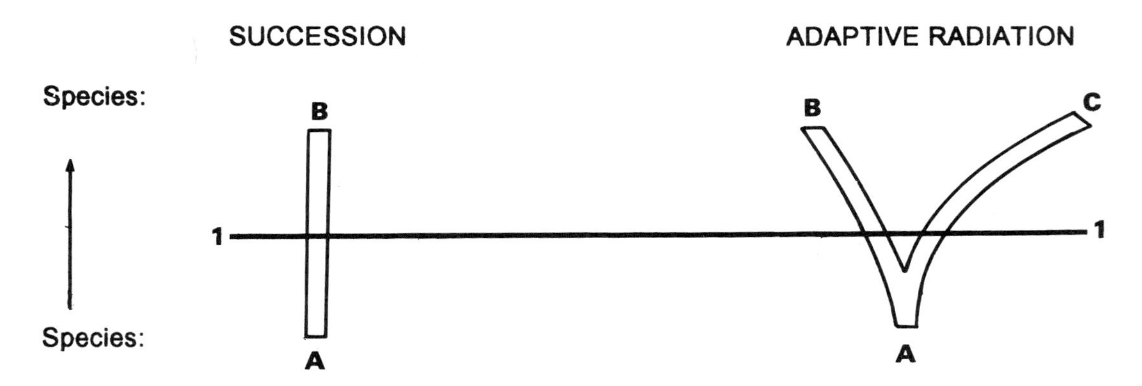

If these species populations are represented only by, for example, skull fragments, and if intervening fossils are found at time 1, considerable disagreement can occur in classifying them. The morphological difference/similarity of the intervening fossil in the single lineage with respect to A and B may lead to lumping it with one or the other species (a subspecies of A or B) or may lead to classifying it as a separate species. The differences/similarities involved in the radiating lineages can evoke even more controversy: Do the two fossils represent isolating subspecies of A, not yet separated as species? Are they sufficiently like B and C to be classified as these species? If both appear morphologically closer to B, does the split into C occur later?

Much of the professional judgment regarding these issues depends on the particular morphological characteristics used as criteria, and this likewise often involves dispute. Not all characteristics -- including skeletal features -- change at uniform rates; not all such traits have underlying genotypes; and thus not all are equally valid as independent criteria for species distinction.

* **Succession** is also known as **Phyletic Evolution** and **Anagenesis**; while **Adaptive Radiation** is also called **Cladogenesis**, referring to branches or **clades**. Darwin's evolutionary concept focused on the former. Lamarck was the first to suggest evolutionary branching.

Assignment: Exercise 8 on the diskette introduces you to some of these problems and allows you to resolve them. In this exercise, you are a Science Officer on the starship Enterprise III and have encountered life forms in a distant galaxy. You must classify these forms and reconstruct their phylogenetic relationships, utilizing observation and the evidence you collect.

There are numerous possible phylogenetic relationships. There is no "correct" one. Use good judgment, however, and make certain that your reconstruction is internally consistent and logical. As 3rd Science Officer, you will be asked specific questions as you proceed. You may even be promoted to Deputy Science Officer.

The problems and decisions you will confront are similar to those encountered by earth-bound Anthropologists working with human fossils. In both cases, there are underlying assumptions which initially guide us and which determine, among other things, whether we tend to be "lumpers" (reluctant to interpret variation as indicating separate species) or "splitters" (tending to see variations as speciation-events).

One such assumption regards the fundamental nature of speciation itself: whether speciation occurs slowly, through the gradual but continual accumulation of genetic change, or more suddenly, in which long periods of genetic equilibrium are punctuated by short-term rapid genetic change. The assumption of **gradualism** implies that intermediate forms will occur and be identifiable throughout the fossil record; the assumption of **punctuated equilibrium** implies that variations will remain relatively constant through time, and a new species will not be preceded by intermediate morphology in the fossil record.

Animal species with records supporting both assumptions exist. Whether the human fossil record supports one or the other assumption depends to a great extent on how the fossils are classified and what criteria are used in the classification.

Exercise 9
Comparative Hominid Crania

Objective: This exercise examines basic features of hominid skulls, in profile, to illustrate some of the major differences in facial and cranial form as these are reflected in the fossil evidence. Since the profiles may be overlaid using a common reference point, you will be able to make direct comparisons in a way not possible in side-to-side comparisons of casts or illustrations. This exercise assumes that you have already become somewhat familiar with the major hominid taxons and their representatives.

Discussion: This is an open-ended exercise which will be facilitated by lecture and/or laboratory materials in your course. When you choose "Comparative Hominid Crania" from the Menu, you will see at the bottom of your screen abbreviations for eight skulls, and the function keys <F1> through <F8> which, when pressed, will draw them. <F9> clears the screen and <F10> returns you to the Menu. The list you see looks like this:

```
F1=Chimp    F2=A.afar    F3=A.af    F4=A.b    F5=H.h    F6=H.e    F7=N    F8=H.s   F9=CLS   F10=EXIT
```

The abbreviations represent species names, except for F7=N, which represents Neanderthal. The individual outlines are presented on page 53 in left-to-right and top-to-bottom order. The abbreviations on your screen represent, respectively, Chimpanzee, Australopithecus afarensis, Australopithecus africanus, Australopithecus boisei, Homo habilis, Homo erectus, Neanderthal, and Homo sapiens sapiens.

The outlines are all drawn to the same scale. They are oriented, for overlay comparison, on the Frankfort Horizontal Plane (horizontal line) and on a perpendicular line at porion (the superior margin of the external auditory meatus, or ear canal). This orientation, however, is for convenience only. Evolutionary changes in the skull did not uniformly proceed from a single point.

You may proceed from one outline to any other in two ways:

1) If you want a direct comparison, you may overlay a second selection on the first without erasing it by simply pressing the respective function key. For example, if you wish to compare the Chimpanzee with Australopithecus afarensis, you will first press F1 and then press F2. You can overlay as many outlines as you wish, in any order. Overlaying several, however, can cause confusion in distinguishing the individual outlines if you are not using an EGA/VGA monitor.

2) If you want to select a skull separately, first press F9. This clears the screen of any drawing(s) currently displayed. Then make your selection.

The illustrated outlines represent specific individual crania. These are as follows: 1) Chimpanzee (SMU Collection); 2) *A. afarensis* (reconstruction from Hadar); 3) *A. africanus* (from Sterkfontein); 4) *A. boisei* ("Zinjanthropus" from Olduvai); 5) *H. habilis* (KNM-ER 1470 from Turkana); 6) *H. erectus* ("Pithecanthropus IV" from Java); 7) Neanderthal ("La Ferrassie" from France); 8) *H. sapiens sapiens* (SMU Collection).

Each cranium bears its own distinct characteristics, and others of the same species will vary, in some cases quite a lot. Moreover, in most of these species there is considerable sexual dimorphism, with males being larger with more robust features than females. Bear this in mind as you use this series, and compare with materials in your text or laboratory.

The drawings are accurate as to contours, but are simplified: The orbits and the zygomatic bones are omitted as is the dentition, except for the occlusal plane. This facilitates the overlays. The saggital crest has likewise been omitted on *A. boisei*; the Chimpanzee, although a male, had none.

The facial bulges on *A. africanus* and *A. boisei* just below the Frankfort Plane are not nasal profiles. They are the anterior thrusts of the zygomatic bone which, because of their forward projections, obscure the otherwise concave upper facial profiles of both species. Many representatives of *A. africanus*, and some of *A. boisei*, have zygomatic bones which do not thrust this far forward.

Please note: On some monitors, these skull outlines may be distorted due to differences in the proportionate height to width ratio of the screen. If there is obvious distortion (such as on most laptop computers), you may wish to switch to a more standard system.

Assignment: Use the screen drawings to compare and contrast some of the more general (not specific) characteristics of these species.

1) Make an initial comparison: Overlay the Chimpanzee and *H. sapiens sapiens*. Compare the general architecture of these skulls. For example, you might mentally divide the lateral views into sections: facial and cranial; upper facial and mandibular; anterior cranial and posterior cranial. Then you might compare these dimensionally: facial height v. anterio-posterior length; cranial height v. length; etc. Where do you see the most profound differences? Recognizing that the chimp and human are greatly different in overall body size, which differences are most likely related to this? For example, if the Chimp were our size, which differences would be reduced in magnitude?

2) You are tempted to view the comparison above as representing the two extremes of our evolutionary progress: The Chimp as the beginning and us as the end. This is untrue: the contemporary Chimp has a separate evolutionary history as long as the contemporary *H. sapiens*. Now make a comparison which does represent these extremes,at least to a high degree: Overlay the *Australopithecus afarensis* and the *H. sapiens sapiens*. These are both hominids, separated by 4 million years or so. Make the same comparisons. Body size differences are still large: *A. afarensis* had a stature of just over about 4 ft., weighing 80-100

lbs. Note that the flattened face and absence of brow-ridges in the *H. sapiens* outline is not universal: some populations of our species have more prognathous faces and slight brow-ridge development.

3) Compare the outlines of *A. africanus* and *A. boisei.* Mentally remove the heavy brow-ridge of the latter and imagine the contour of the frontal bone. Where are the most significant differences? What are the most significant differences in the mandibles of the two? How are these reflected in the dentition?

4) Homo habilis is an important species in many ways: it represents a transition from the smaller-brained Australopithecines to the larger-brained *H. erectus,* and its contemporaneity with *A. boisei* suggests a separate evolutionary lineage for the latter. Compare *H. habilis* with the gracile *A. africanus.* Point out two major cranial differences. Examine the facial differences and similarities. Some gracile forms in E. Africa suggest more similarities to the gracile S. African than to habilis. Unfortunately, more complete facial evidence for *H. habilis* is ambiguous at present. How does *H. habilis* contrast with boisei?

5) Overlay *H. habilis* with *H. erectus.* In what two ways do their facial profiles differ? Compare the total facial prognathism of habilis with the lower facial prognathism of *H. erectus.* Other examples of *H. erectus,* from other regions, have crania which are shorter in anterio-posterior length and with greater vault height (porion-bregma). This Java specimen is perhaps the most primitive, with cranial capacity just over 700 cc. Others range just beyond 1000 cc. Given the current appearance of the *H. habilis* face and cranial base, as seen in the outline, do you think the jaw would be more like the Australopithecines or erectus in its several dimensions? The ascending mandibular ramus of erectus tends to be square in most erectus representatives. Is it probably more rectangular in habilis or more nearly square?

6) "Neanderthal" is a term which encompasses the "Classic" forms in SW Europe and more varied forms elsewhere (the map in Exercise 11 shows these locations). All are *H. sapiens.* This example is one of the "Classic Neanderthals". What are the distinguishing features of the "Classics", and how many can you identify in the skull outline? Compare this Neanderthal with *H. erectus.* In which regions of the skull and face are they roughly similar and where are they different? Now overlay *H. sapiens sapiens* and make the same comparisons. This latter is a contemporary example. What is the time difference between them?

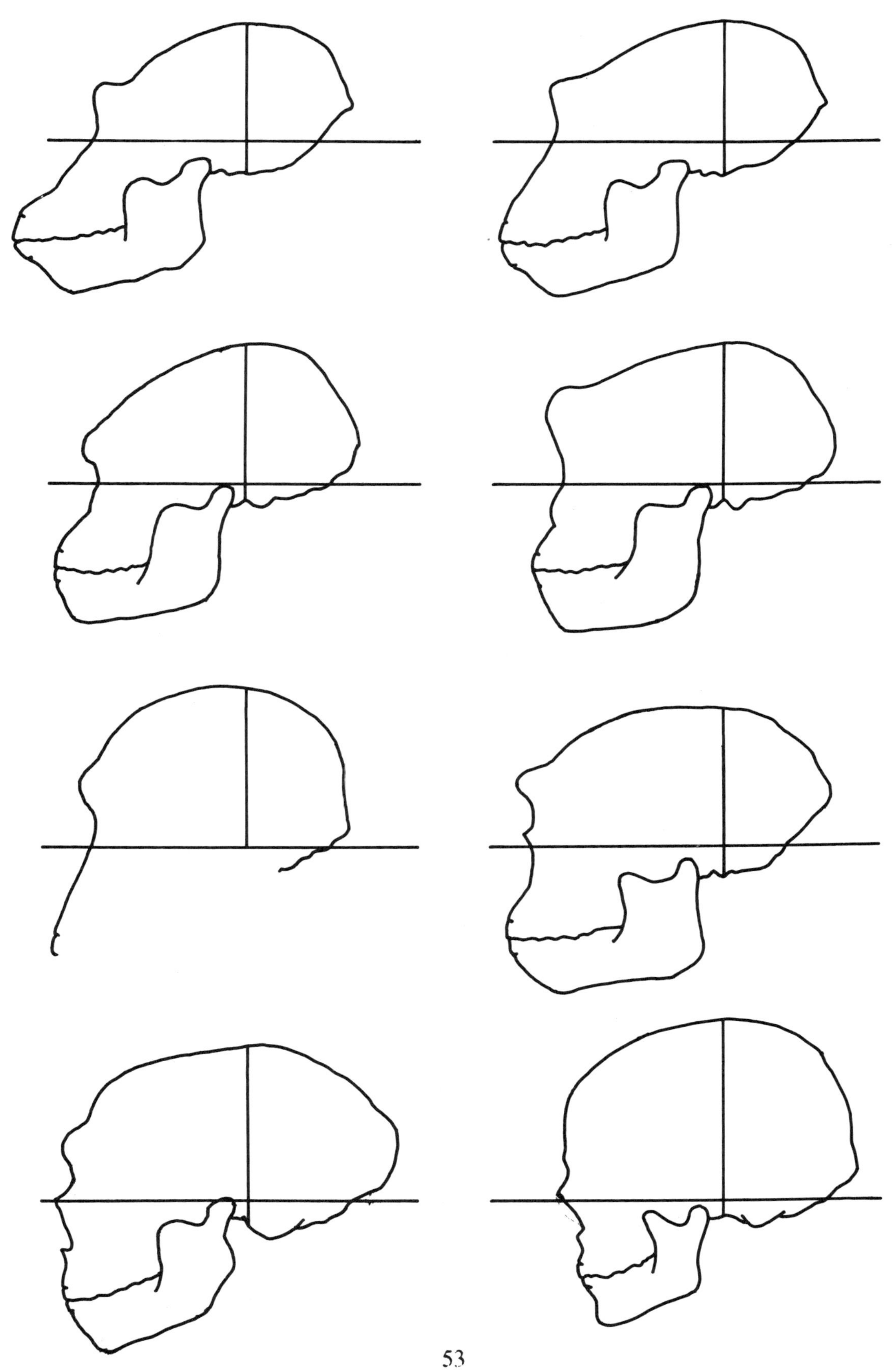

Exercise 10
East African Hominid Distributions

Objective: In this exercise you will compare site locations and time periods in East Africa, and the species assignments they are most commonly given. You should use text/lecture materials for more detailed information and comparisons.

Discussion: The countries of Ethiopia, Kenya, and Tanzania together have provided us with not only the largest inventory of hominid fossils of any like region of the world, but also the best dated. The earliest hominid forms come from here (and, according to some scholars, the founding members of our own species, Homo sapiens).

Hominid recoveries in general, but particularly in this region, occur with such regularity that published textual discussions are often incomplete or obsolete even before they reach your bookstore. Controversies concerning hominid evolutionary interpretations are likewise continually changing, with arguments often hinging on esoteric details of morphology.

Consequently, it is not possible to affirm "the latest information" available as this is written to be the latest information available when you read it. The hominid distributions plotted are some (not all) of the current and most important representatives of the early forms of humanity: the several australopithecine species and the initial species of our genus, Homo.

Assignment: From the Hominid Distribution Maps on your MENU, choose "East Africa". The numbered selections in the menu box designate your choices. The last, (5 = Features), when pressed, identifies the countries and topographic features. Each of the remaining selections plot the locations and identify fossils, or their site proveniences, which represent a particular hominid species as most commonly accepted. You are asked, in each case, to name this species. After you name the species (you may abbreviate) and press the <Return> key, the correct species (or current consensus) is given. The last option exits the program and returns you to the MENU.

Your instructor and/or text may add additional information, illustration, or interpretation to these. If you have a laboratory with the course, you may examine casts of some of these fossils and cover their morphologies in some detail.

The materials shown on the map are listed and briefly discussed here by locality, rather than by species, so that you may have some perspective on the time sequences and richness of information represented at specific sites.

ETHIOPIA

Hadar sites, in the Afar Triangle, approx 150 mi. NE of Addis Ababa.
Dated: 3.3-3.0 million years ago (mya). Numerous human male & female specimens, including a partial skeleton, dubbed "Lucy". *Australopithecus afarensis*.

Belohdelie, middle Awash R. valley; fragmentary frontal bone dated more than 4 mya. Probably *A. afarensis*.

Maka, middle Awash R. valley; upper left femur fragment, dated 3.5-4 mya. Probably *A. afarensis*.

Omo Valley, along the Omo River north of Lake Turkana. Dated: 3.0-1.1 mya. A stratigraphic sequence revealing gracile Australopithecines similar to *A. africanus* (3.0-2.5 mya); *A. boisei* (2.5-1.0); Homo habilis (1.85 mya); and *H. erectus* at about 1 mya.

KENYA

Lake Turkana. On the West side, dated 2.5 mya, the "Black Skull" (WT 17000), *A. boisei*; and, dated 1.6 mya, WT 15000, *H. erectus*.
On the East side, at Koobi Fora, dated 2.2-1.5 mya, *H. habilis* (ER 1470; ER 1813), and A. boisei (ER 406, ER 703, ER 732). Dated 1.6-1.5 mya, *H. erectus* (ER 3733; ER 3833). Numerous fragments and post-cranial remains, as well as stone tools.

Kanapoi. South side of Lake Turkaka; distal humerus fragment, dated more than 4 mya. Possibly *A. afarensis*.

Lothagam. West of Lake Turkana; mandible fragment, dated 5.5 mya. Possibly *A. afarensis*.

TANZANIA

Olduvai Gorge. Bed I (1.8-1.65 mya), the "Zinjanthropus" *A. boisei* skull (OH 5), and *H. habilis* (OH 7, OH 24); Bed II (1.6-1.2 mya), *H. habilis* (OH 13) and *H. erectus* (OH 9); Bed IV (600,000-400,000 years ago), *H. erectus*.

Peninj. Lake Natron, dated about 1.5 mya, a single mandible. *A. boisei*.

Laetoli. South of Olduvai Gorge, date: 3.75-3.6 mya. Numerous upper and lower jaw fragments from discoveries made over the years from 1945 to the 1970s. Also preserved footprints in hardened volcanic tuff. *A. afarensis*.

Exercise 11
North African and European Hominid Distributions

Objective: In this exercise you will compare site locations and time periods in North Africa and Europe, and the species assignments they are most commonly given. You should use text/lecture materials for more detailed information and comparisons.

Discussion: This region, particularly Europe, has a most complex fossil history, even though only two hominid species, Homo erectus and *H. sapiens* (including the Neanderthals), are known to have occupied the European continent. This occupation covers that portion of the Pleistocene during which major glaciations occurred. The Mindel (Elster, in some texts), Riss (Saale), and Wurm (Weichsel) polar ice sheets descended, in turn, to cover much of the European land mass.

At the same time, the volume of water locked into glacial ice caused a lowering of sea level, exposing new land along sea coasts and opening land bridges to connect otherwise separate land bodies. During **interglacial** periods, ice masses retreated and sea levels once again rose. This repeated sequence not only altered the terrain and the climate, but changed habitats for plants and animals as well.

It is in this context that our ancestors developed new strategies for survival, in large part responsible for the evolutionary changes which eventually resulted in our modern morphology and physiology. To understand these evolutionary changes, and how they relate to the movements of people across this vast landscape through time, we examine the fossil and the archaeological remains they left behind.

With some important exceptions, the fossil evidence is quite spotty, reflecting single specimens rather than populations. Accuracy of dating is even less certain: most dates come from the association of fossils with geological strata or faunal remains for which only relatively broad estimated dates may be assigned.

Two uncertain factors thus influence the identification of evolutionary change: the interpreted date and the interpreted morphology. These are the prime ingredients for controversy, and controversy has characterized the European fossil scene since the initial Neanderthal discovery in 1856.

You will use this exercise as a resource to guide you in your course. As you discuss particular time periods or specific fossils and sites, the maps and information here should provide you with a better understanding of their spatial relationships. Understand that not all fossils or site localities are shown. A majority of those most commonly discussed are, however. Your instructor will guide you in identifying which are important to your course and which are not.

Assignment: From the MENU for this Exercise Set, choose "Europe". You have two options:

First, you may choose one of the broad time periods available. That selection will show the distribution of fossils and/or sites most commonly assigned to that period. The map will also show, where appropriate, the extent of glacial ice and the altered coast line. The time periods are: 1) Mindel Glaciation, 2) Mindel/Riss Interglacial, 3) Riss Glaciation, 4) Riss/Wurm Interglacial, and 5) Wurm I - Wurm II Glaciations. The last of these combines fossils from the first two Wurm Ice advances (**stadials**) and the intervening Wurm I/II **interstadial** (a brief interglacial period between these).

Second, you may choose "Select Fossil/Site". This selection will provide you with a list of the fossils or sites shown below. Choose by its number and press <Return>. You will be asked to name its country and its time period (from a list of time periods). Press <Return> after each entry. The fossil will be plotted on the map, its correct country and time period given, and a brief description of it will be provided.

You may use this second option as a test to check your knowledge, or simply to get information. (You may press <Return> without entering anything and still receive the information). You may want to use the list below to check off those fossils or sites relevant to your course.

The fossils/sites included are:

___ Arago
___ Biache
___ Bilzingsleben
___ La Chaise
___ Classic Neanderthals
___ Ehringsdorf
___ Fontèchevade
___ Gibraltar
___ Jebel Ihroud
___ Krapina
___ Lazaret
___ Mauer
___ Monte Circeo
___ Montmourin
___ Petralona
___ Rabat
___ Saccopastore
___ Salé
___ Spy
___ Steinheim
___ Swanscombe
___ Ternifine (Tighenif)
___ Terra Amata
___ Thomas Quarry
___ Torralba/Ambrona
___ Vértesszöllös

Exercise 12
Asian Hominid Distributions

Objective: In this exercise you will compare site locations and time periods in Southern and Southeast Asia, and the species assignments they are most commonly given. You should use text/lecture materials for more detailed information and comparisons.

Discussion: The Asian subcontinent is the region of some of the earliest and most remarkable discoveries of human fossils ever revealed. Representative fossils of the species *Homo erectus* were first discovered here, and provided the initial paleontological support for Darwin's assertion that human and ape ancestry stemmed from a common lineage. Many of the controversies surrounding these early discoveries have not yet been completely resolved.

In 1891, the first of these discoveries was made in Java by the Dutch physician Eugene Dubois, who went to Southeast Asia with the explicit intent to find Darwin's "missing link". His discovery was made along the Solo River near the village of Trinil, and he triumphantly assigned it the name Pithecanthropus erectus ("erect ape-man"). The remains consisted of a primitive skullcap and a femur which evidenced bipedalism.

Similar circumstances surround the discovery of *H. erectus* in China ("Peking Man") in 1927. The Canadian physician Davidson Black, like Dubois, was convinced that a human ancestor lay waiting for him to discover. Excavating at a cave site on Dragon Bone Hill, near the village of Zhoukoudian (previously Choukoutien), Black and Chinese colleague W. C. Pei discovered fossil fragments he named Sinanthropus pekinensis. Massive excavations over the following ten years revealed 14 skulls and over 150 teeth. Subsequent study and casts made by Franz Weidenreich carefully documented these materials before they became lost during World War II.

These, and many other discoveries since, have demonstrated that *Homo erectus* occupied southern and southeast Asia from early Pleistocene times. How did *H. erectus* get here? Did the species evolve locally from a preceding form? Despite some problematic fossils suggesting an Australopithecine ancestor, no earlier species has been satisfactorily identified in this region. Authorities now accept the early expansion of *H. erectus* out of Africa and into Asia as the most likely source.

Many of these Asian examples show morphological characteristics suggesting regional adaptations distinct from *H. erectus* remains elsewhere. Furthermore, remains of archaic *Homo sapiens* in Asia -- during the period when Neanderthal characteristics were common in Europe and elsewhere -- suggest continuity in these regional traits, cutting across the erectus-sapiens species boundary.

This broad inventory of Asian fossils provides the strongest support for the "Multiregional Hypothesis" first proposed by Weidenreich, and championed by CarletonCoon and C. Loring Brace: that *H. erectus* evolved into *H. sapiens* in several regions of the Old World during the Middle Pleistocene. This hypothesis, elaborated by Milford Wolpoff, challenges the recent "African Origins" ("Garden of Eden") Hypothesis of Rebecca Cann, Christopher Stringer, and others, which claims that anatomically modern *H. sapiens* evolved in Africa and thence spread into all other regions, replacing pre-existing humans.

One of the major centers of controversy between these two hypotheses lies in the interpretation of the Asian material. This exercise provides data on many of the fossils involved in these issues. Like the preceding exercise on the North African and European material, it consists of a series of maps showing hominid locations during the Pleistocene, and offers brief descriptions of each fossil after quizzing you on locations and time periods. It differs in two major respects from the previous exercise:

First, there are no glacial sequences here. Ice sheets during the Pleistocene never reached into this region. However, during continental glaciation elsewhere the sea level was lowered, and the maps indicate the contiguous land exposed (at a 600-ft. depth) during these episodes. This will show when land bridges connected the islands of S.E. Asia to the mainland.

The second difference is in chronology. The absence of glacial events in this region, plus the absence of volcanic formations capable of more precise radiometric dating, makes the dating of fossil sites much more problematic. Hence, the ranges of dates provided for this material is in some cases quite debatable. Estimates of dates for some materials have changed remarkably in recent years, and you may find some disagreement between those suggested here and those in a particular text. Your instructor may correct or help resolve disputed dates. For comparison, each time period designated the approximate equivalence to the named Alpine glacial stages in Europe.

Assignment: From the MENU for this Exercise Set, choose "Asia". You have two options:

First, you may choose one of the broad time periods available. That selection will show the distribution of fossils most recently assigned to that period. The map will also show, where appropriate, the altered coast line and exposed land mass. The time periods, and the approximate time ranges for the included fossils, are: 1) Lower Pleistocene (1 to .65 million years ago); 2) Early Mid-Pleistocene (650,000 to 300,000 yrs ago); 3) Middle Pleistocene (300,000 to 200,000 yrs ago); and 4) Late Middle Pleistocene (200,000 to 120,000 yrs ago).

Second, you may choose "Select Fossil/Site". This selection will provide you with a list of the fossils shown below. Choose by its number and press <Return>. You will be asked to name its country and its time period (from a list of time periods). Press <Return> after each entry. The fossil will be plotted on the map, its correct country and time period given, and a brief description of it will be provided.

You may use this second option as a test to check your knowledge, or simply to get information. (You may press <Return> without entering anything and still receive the information). You may want to use the list below to check off those fossils relevant to your course.

The fossils included are:

___	Changyang	___	Dali	___	Hexian
___	Jinniushan	___	Lantian	___	Maba
___	Narmada	___	Perning (Modjokerto)	___	Sambungmacan
___	Sangiran	___	Solo (Ngandong)	___	Trinil
___	Zhoukoudian				

Exercise 13
Reconstructing Human Phylogeny

Objective: In this final exercise you will draw upon the evidence you have studied to reconstruct the hominid phylogenetic sequence, using specific fossils. At this point, you should have completed your examination of the fossil evidence. If you have not already completed Exercise 8 (Phylogenetic Principles), you should do so first.

Discussion: The reconstruction of human phylogeny, and its representation in the form of an ancestral (phyletic) "tree", with its lineages and branches representing succession and adaptive radiation, is an intriguing exercise in human paleontology. Such "trees" attempt to show the evolutionary relationships among evolving species using particular fossils to represent populations or lineages.

If, in running Exercise 8, your philosophy was that of an extreme "lumper", you placed all of the Critters in a single (**monophyletic**) lineage: your "tree" had no branches. If your philosophy was that of an extreme "splitter", each Critter form defined its own branch, and your "tree" was **polyphyletic**.

Both philosophies, as fundamental **presuppositions** about the nature of species evolution, have characterized approaches to human phylogeny and have played important roles in defining basic models of early hominid evolution.

At one extreme, the "single species" hypothesis has claimed that no more than one hominid species was ever present at one time. Supported by the **competitive exclusion principle** in ecological science (competitors in the same niche cannot coexist), advocates have asserted that the broad adaptive zone (alternative habitats) provided by initial cultural ability made it impossible for two species of hominids to have survived as contemporaries. Hence, human phylogeny consists of successive species in a single lineage, with morphological variation distinguishing populations as the **subspecies** level.

At the other extreme, the polyphyletic hypothesis has claimed that several coexisting hominid species, particularly in the earliest phases of our evolution, represent a well-established principle: Upon the emergence of a new level of biological organization (**grade**), adaptive radiation rapidly occurs, and the new level diversifies into many species, often adapting to habitats which differ in few details. The primate record itself reveals such radiations from the initial Prosimians in the Paleocene to the Dryopithecines of the Miocene. Hence, human phylogeny consists of several coexisting species in each grade, with only one surviving to evolve into the next grade.

As in most of science, the evidence tends to support elements of both extremes. The alternative phyletic reconstructions common in the literature reflect differing combinations of these philosophies, often without any single underlying paradigm. Of course, such reconstructions are always changing as new evidence emerges, and no modelis likely to go unchallenged. After all, it is an interpretation of reality -- not reality itself -- that such models reflect.

Assignment: In this exercise, you will use your own judgment to create a phyletic tree of hominid evolution. The information you use will be drawn from the evidence. You, however, will decide which evidence to apply.

The basic problem is to apply criteria which help to identify the process of **sapienization**. Fossil representatives will be compared on a less-to-more sapiens-like scale: is **A** closer to sapiens than **B** is? If so, is **C** more sapiens-like than **B**? And so on.

In making these comparisons, you are given a list of six hominids:

1. Australopithecus afarensis
2. Australopithecus africanus
3. Australopithecus robustus
4. Australopithecus boisei
5. Homo habilis
6. Homo erectus

You may add, if you wish, an additional name (for example, some specific fossil). You will list these in their descending order of sapienization, or membership in the sapiens lineage: choose the most sapiens-like first, followed by the next-closest phyletically related fossil. List the least sapiens-related fossils last (in order of their own lineage relationships, if you choose a multi-clade model).

For example, if you assume a single-lineage model, all fossils will be related to one another in one line, with the least sapiens-like member listed last. If, however, you assume that *A. boisei* (for example) is in a separate lineage with *A. afarensis* a common ancestral species to both boisei and sapiens, the latter will be listed before boisei because it is in our lineage, even though afarensis is less "sapiens-like" than boisei. You may need to construct several models before you arrive at the sequence that reflects your particular assumptions.

After each selection, you are asked to give its earliest date in millions of years before present (e.g., 1.3, 2.5, 4, etc.). This will place the form correctly on the chart. Allowable dates range from 1 to 5.5 MYBP. When you press <Return> the final time (when the cursor is at the 7th form, or, if you add a 7th, when you have entered its age) your diagram will be presented in graphic form, as illustrated in the examples below, but without their connecting lines.

You may save this phyletic tree as an option (it will be called MYCHART) and print it on your printer. Using the printed copy, you then draw the lines connecting the fossils according to your judgment.

When you run this exercise on the diskette, you are first asked if you wish to view MYCHART from your previous "save". If you have not already constructed one, you should enter "N" for "No".

Discussion: You will note that the "sapienization" scale is not numerical: the increments from left to right have no value labels. This is an arbitrary scale and each unit, as you proceed to the right, is an increase in sapienization in the same degree. Your phyletic tree simply reflects the judgment that each hominid to the right is closer to sapiens than the one preceding it, on the basis of whatever criteria you choose.

Given this sequence, it is up to you to complete the tree by connecting those hominids which in your judgment stand in direct evolutionary relationship with one another. Scientists make these judgments based not only on the degree of shared characteristics, but also on 1) the magnitude of the change in those features used as criteria, 2) the amount of time which separates them, and 3) whether the time difference appears to be adequate to allow the change.

Two alternative models are illustrated below, drawn from this program. Disputes arise not only over the criteria used for phylogenetic reconstruction but over the judgments used in applying these criteria. Compare your own reconstruction with others to identify the differences in the criteria and in their application. As in many of these exercises, there is no "correct" solution. Feasibility rests on your argument from the evidence. There are, of course, numerous <u>incorrect</u> solutions!

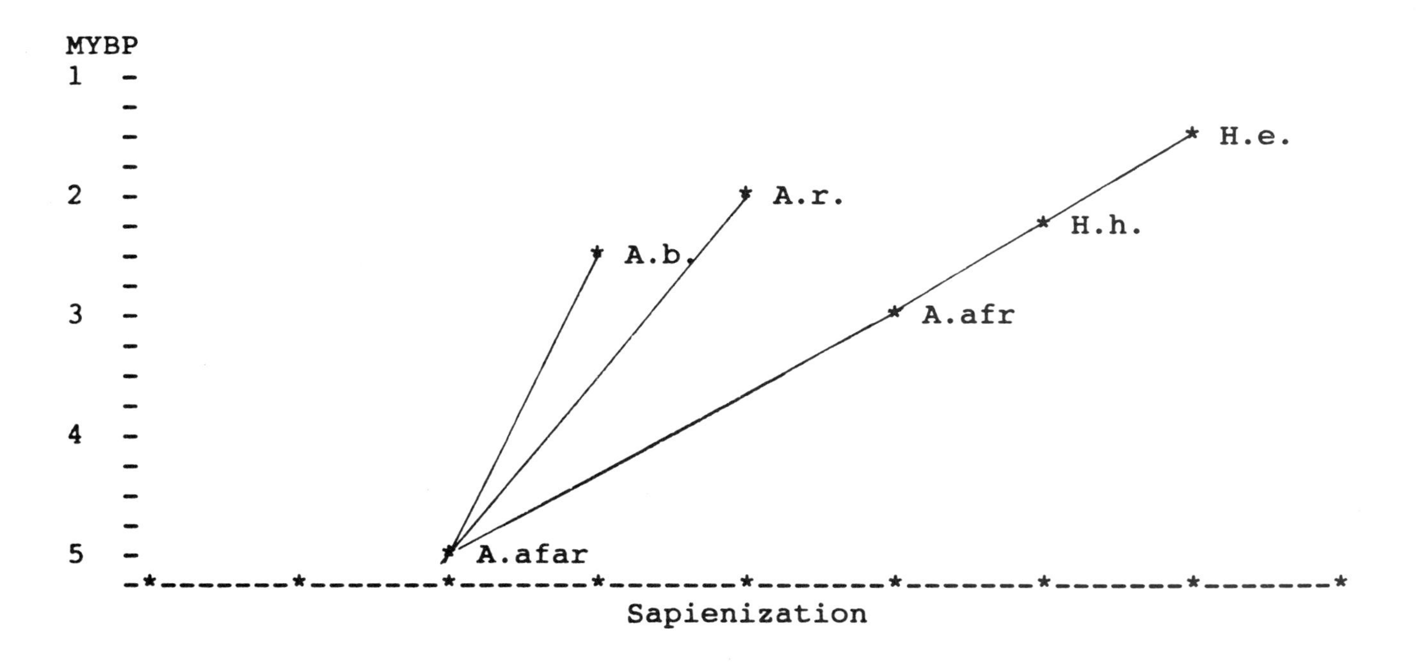

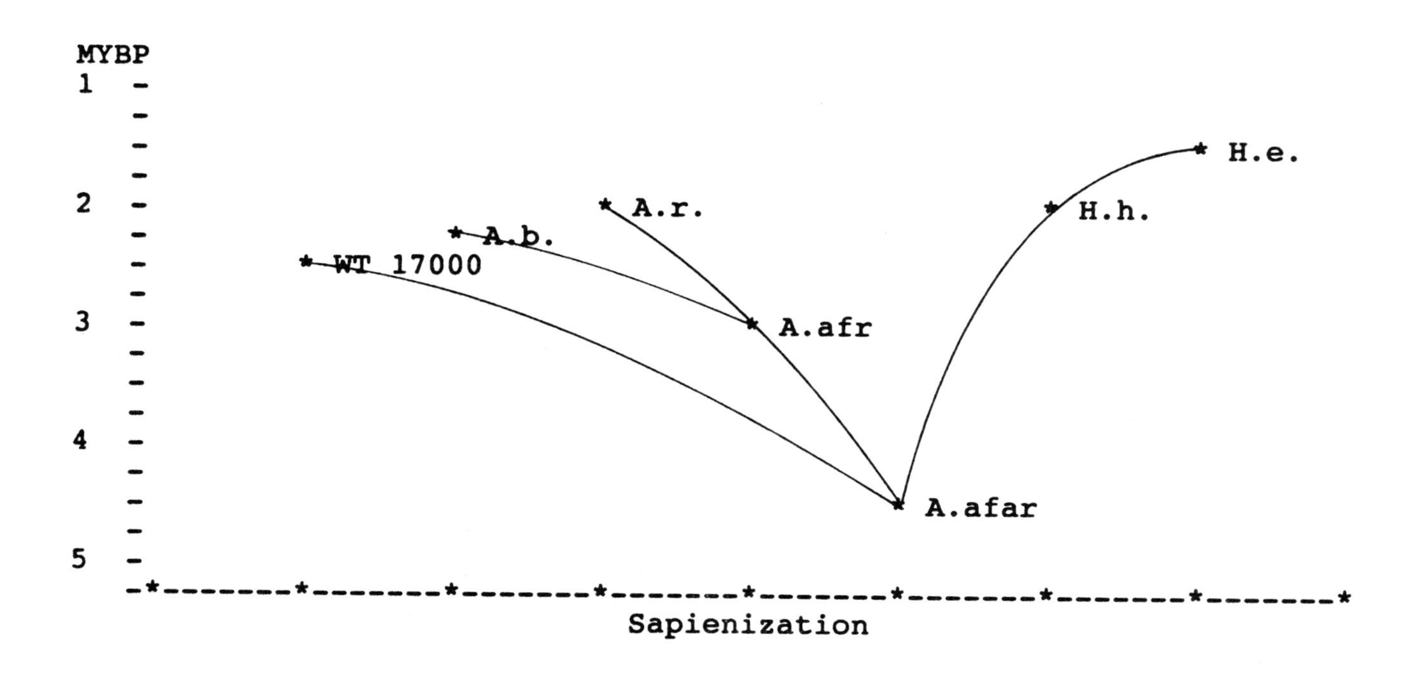
MYBP
1
2
3
4
5
H.e.
A.r.
H.h.
A.b.
WT 17000
A.afr
A.afar
Sapienization

APPENDIX

Quizzes on Exercises 1-5

The following quizzes cover the concepts central to the first five exercises on Evolutionary Genetics. These are the concepts which lie at the core of our understanding in evolution, but frequently are difficult for students to grasp.

These quizzes test your understanding of terms, concepts, and relationships. They are not quizzes on mathematics (although some math may be used as illustration). The principles behind the math are the focus of these quizzes.

Quiz 1: Mendelian Genetics (Exercises 1-2)

1. The term "gene" is most nearly equivalent to the term
   a. allele b. chromosome c. locus d. gamete

2. The genetic unit ("gene") which determines a particular trait, such as blood type, occurs at a particular point on a
   a. chromosome b. blood cell c. fertilized egg d. gamete

3. The particular point where a gene occurs is known as
   a. a genotype b. an allele c. a locus d. a chromosome

4. Chromosomes exist in pairs in all normal cells except
   a. homozygotes b. heterozygotes c. zygotes d. gametes

5. Gene loci exist in pairs in
   a. somatic cells b. reproductive cells c. gametes d. egg cells

6. An allele is an alternative
   a. expression of a particular genotype
   b. expression of a particular phenotype
   c. form of a chromosome in a particular pair
   d. form of a gene at a particular locus

7. Together, the pair of alleles at a paired locus describes that trait's
   a. phenotype b. genotype c. homozygote d. heterozygote

8. The manifested characteristic determined by a pair of alleles is the
   a. phenotype b. genotype c. homozygote d. heterozygote

9. If the two alleles at a paired locus are the same in an individual, the genotype of that locus is
   a. diploid b. haploid c. homozygous d. heterozygous

10. If one of the two alleles at a locus expresses itself at the expense of the other, the former allele is said to be
   a. dominant b. recessive c. co-dominant d. superior

11. A heterozygous genotype exists at a locus if
   a. the paired alleles are the same
   b. the paired alleles are different
   c. the paired alleles are both dominant
   d. the paired alleles are both recessive

12. When a recessive condition is expressed in the phenotype, the genotype is
   a. diploid   b. haploid   c. heterozygous   d. homozygous

13. At a heterozygous locus in which <u>each</u> allele expresses itself in the phenotype, the two alleles are said to be
   a. dominant   b. recessive   c. co-dominant   d. heterozygous

14. Meiosis is cell division which results in
   a. normal tissue growth in the body
   b. the production of diploid (2n) cells
   c. the production of haploid (1n) cells
   d. a fertilized egg (zygote)

15. When a heterozygous locus undergoes meiosis, the result is
   a. two different kinds of gametes, each containing one of the alleles
   b. one kind of gamete, containing both of the alleles
   c. two different kinds of gametes, containing either both alleles or none
   d. one kind of gamete, containing the only allele at that locus

16. Segregation refers to the separation of
   a. different loci during meiosis
   b. alleles at a paired locus during meiosis
   c. different gametes during fertilization
   d. different genotypes among offspring

17. If two heterozygous loci in an individual produce only two kinds of gametes,
   a. segregation is responsible
   b. independent assortment is responsible
   c. both segregation and independent assortment are responsible
   d. neither segregation nor independent assortment are responsible

18. If two heterozygous loci in an individual produce four kinds of gametes
   a. segregation is responsible
   b. independent assortment is responsible
   c. both segregation and independent assortment are responsible
   d. neither segregation nor independent assortment are responsible

19. For a gene locus with three possible alleles (e.g., the ABO-locus), a person
   a. can be heterozygous and have all three alleles
   b. can only be homozygous, possessing a pair of one of the three
   c. can only be heterozygous, possessing any two of the three
   d. can be homozygous for one of the three, or heterozygous for any two of the three

20. Combining the Mendelian principles of segregation and independent assortment, a person who is heterozygous at n loci, each on a different chromosome pair, can produce how may different kinds of gametes?
   a. $n \times 2$ b. $n \times 4$ c. 2^n d. n^2

Quiz 2: Population Genetics (Exercise 3)

1. Consider a population of 400 people. At the ABO-locus the number of people with each genotype is as follows:

Genotypes:	100	50	50	100	100
No. of people:	AA	BB	AB	AO	OO

a. How many people are homozygous for allele A?
b. How many people are recessive homozygotes?
c. How many people in all have homozygous genotypes?
d. How many people have type-A blood?
e. How many type-A persons are heterozygotes?

2. For the population above, give the following frequencies in decimals:
a. What is the frequency of the homozygous A genotype?
b. What is the frequency of type-O individuals?
c. What is the frequency of type-A individuals?

3. For a population in Hardy-Weinberg equilibrium at the pq-locus, the frequency of allele **p** is
a. p^2 b. p^2 c. $p^2 + pq$ d. both b. and c.

4. For a population not in Hardy-Weinberg equilibrium at the pq-locus, the frequency of allele **p** is
a. p^2 b. p^2 c. $p^2 + pq$ d. both b. and c.

5. For any population, at the pq-locus
a. $p + q = 1$ b. $p^2 + q^2 = 1$ c. $p^2 + q^2 = 1$ d. both a. and b.

6. For a population in Hardy-Weinberg equilibrium
a. $p + q = 1$ b. $p^2 + q^2 = 1$ c. $p^2 + q^2 = 1$ d. both a. and b.

7. At the PTC-taster locus (T-t), the dominant allele designates a taster. In a population of 1000 persons, the genotypes are:

Genotypes:	TT	Tt	tt
No. of people:	640	200	160

(Using decimals for all frequencies):
a. What is the frequency of non-tasters?
b. What is the frequency of the recessive allele?
c. What is the frequency of the dominant allele?

8. In this population, the genoytpe frequencies at the PTC-locus are:

Genotypes:	TT	Tt	tt
No. of people:	.64	.20	.16

a. What is the frequency of allele T if calculated as $p^2 + pq$?
b. What is the frequency of allele t if calculated as $q^2 + pq$?
c. What is the frequency of allele T if calculated as p^2 ?
d. What is the frequency of allele t if calculated as q^2 ?
e. Is this population in Hardy-Weinberg equilibrium?

9. In each of the following populations, give the allele frequency asked for and designate (Y or N) whether the population is in Hardy-Weinberg equilibrium:

Genotype frequencies:			Allele freq.:	H-W ?:
a. AA = .36	AO = .48	OO = .16	A = _____	____
b. TT = .25	Tt = .50	tt = .25	t = _____	____
c. rr = .01	Rr = .50	RR = .49	r = _____	____
d. MM = .49	MN = .42	NN = .09	M = _____	____
e. PP = .25	PQ = .66	QQ = .09	P = _____	____

10. Assume that the following populations are in Hardy-Weinberg equilibrium for each 2-allele locus listed. Write the frequencies of the designated genotypes, using decimals:

Allele frequencies:		Genotype frequencies:
a. p = .5	q = .5	p^2 = _____
b. T = .6	t = .4	Tt = _____
c. r = .3	R = .7	RR = _____
d. A = .1	a = .9	Aa = _____
e. O = .8	B = .2	OO = _____

11. For the populations above, give the designated phenotype frequencies:

a. Heterozygotes ____

b. Non-tasters ____

c. Rh negatives ____

d. Dominant phenotype ____

e. Blood type-B ____

Quiz 3: Natural Selection (Exercise 4)

1. We can conclude that selection is acting if
   a. one phenotype has a higher mortality rate than another
   b. one phenotype has a lower fertility rate than another
   c. one phenotype has a higher frequency than another
   d. both a. and b. above
   e. all of the above

2. Assume that in a population a locus with two alleles produces, among its three genotypes, only <u>two</u> phenotypes. If one phenotype has a higher mortality rate than the other
   a. that phenotype will increase in frequency each generation
   b. that phenotype will decrease in frequency each generation

3. If the phenotype with the higher mortality rate (above) is produced by two genotypes
   a. selection is operating against a dominant trait
   b. selection is operating against a recessive trait
   c. selection is operating against a co-dominant trait

4. If one phenotype has a higher fertility rate than the other
   a. that phenotype will increase in frequency each generation
   b. that phenotype will decrease in frequency each generation

5. If the phenotype with the higher fertility rate (above) is produced by only one of the genotypes
   a. selection is operating against a dominant trait
   b. selection is operating against a recessive trait
   c. selection is operating against a co-dominant trait

6. If one phenotype has a selection value (coefficient) of 80%
   a. that phenotype may be said to have a fitness value of 20%
   b. the other phenotype may be said to have a fitness value of 20%

7. If selection operates against a dominant phenotype
   a. the dominant allele will eventually reach a frequency of 100%
   b. the recessive allele will eventually reach a frequency of 100%
   c. the dominant allele will be systematically reduced but not eliminated
   d. the recessive allele will be systematically reduced but not eliminated

8. If selection operates against a recessive phenotype
   a. the dominant allele will eventually reach a frequency of 100%
   b. the recessive allele will eventually reach a frequency of 100%
   c. the dominant allele will be systematically reduced but not eliminated
   d. the recessive allele will be systematically reduced but not eliminated

9. Where selection operates against the dominant phenotype, the <u>rate</u> of reduction of the dominant allele will increase
   a. as the number of generations of selection increases
   b. if the mortality rate of the dominant phenotype increases
   c. both of the above

10. The following population is in Hardy-Weinberg equilibrium at the **pq**-locus:

Genotypes:	p^2	2pq	q^2
Frequencies:	.25	.50	.25

One year a disease is introduced, resulting in the following fitness values:

Genotypes:	p^2	2pq	q^2
Fitness:	<1.0	1.0	1.0

Over the generations, we would expect the p^2 genotype to:

a. increase to 100%
b. increase but not reach 100%
c. decrease to 0%
d. decrease but not reach 0%

11. Assume that the fitness values at this locus change to the following:

Genotypes:	p^2	2pq	q^2
Fitness:	<1.0	<1.0	1.0

Over the generations, we would expect the q^2 genotype to:

a. increase to 100%
b. increase but not reach 100%
c. decrease to 0%
d. decrease but not reach 0%

12. Assume that the fitness values at this locus change to the following:

Genotypes:	p^2	2pq	q^2
Fitness:	<1.0	1.0	<1.0

Here, selection operates against both homozygotes. We expect that, over the generations,

a. both homozygotes will eventually be eliminated
b. the homozygote with the lowest fitness will eventually be eliminated
c. the heterozygote and one homozygote will eventually reach equilibrium
d. both homozygotes will eventually reach equilibrium

13. Finally, for this population, assume that the frequencies and fitness values are as follows:

Genotypes:	p^2	2pq	q^2
Frequencies:	.25	.50	.25
Fitness:	1.0	<1.0	1.0

a. Both alleles will eventually be eliminated
b. The heterozygote will eventually be eliminated
c. Both alleles will remain at 50% frequencies
d. Both alleles will be reduced but not eliminated

14. You have been asked about four different ways in which natural selection can occur at a locus with 2 alleles, each depending on the kind of phenotypes which are involved:

A. Selection against a dominant phenotype
B. Selection against a recessive phenotype
C. Selection against both homozygous phenotypes
D. Selection against the heterozygous phenotype

a. In which case will selection maintain genetic variability in a consistent manner?

b. In which case will selection consistently eliminate variation?

Quiz 4: Gene Flow (Exercise 5)

1. Gene flow occurs when
   - a. two or more populations have members who intermarry
   - b. one population migrates to the locality of another
   - c. two populations merge to form a single, hybrid population
   - d. any of the above occurs

2. Gene flow between two populations
   - a. increases the genetic differences between them
   - b. decreases the genetic differences between them
   - c. maintains the genetic differences between them
   - c. has no effect on the genetic differences between them

For questions 3-6:

Populations A and B unite through the migration of one to the other. Hybrid population C equals the combined gene pools of A and B. At the **pq** locus, original Population A had a frequency of p = .9 and Population B had a frequency of p = .1

3. The two populations each had 150 individuals. The proportionate genetic contribution of Population A to Population C is

   a. 15% b. 25% c. 50% d. 100%

4. The two populations each had 50 individuals. The proportionate genetic contribution of Population A to Population C is

   a. 15% b. 25% c. 50% d. 100%

5. Population A had 100 people, while Population B had 300 people. The proportionate contribution of Population A to Population C is

   a. 15% b. 25% c. 50% d. 100%

6. With equal numbers of people in Populations A and B, Population C will have a frequency of p =

   a. 1.0 b. .90 c. .10 d. .50

For questions 7-10:

Population A and Population B routinely intermarry at a rate of 10%. Thus, 10% of all marriages in Population A have one marriage partner from Population B, but half (5%) move from A to B.

7. The gene flow rate for Population A is

   a. 1% b. 5% c. 10% d. 90%

8. If at present Population A has a frequency at the **pq**-locus of p = .9 and Population B has a frequency of p = .1, with 5% gene flow, the next generation of Population A will have
   - a. 95% of its gene pool represented by alleles from Population B
   - b. 50% of its gene pool represented by alleles from Population B
   - b. 5% of its gene pool represented by alleles from Population B

9. The frequency of **p** in Population A after the initial 5% gene flow will be

   a. (.9 x .05) + (.1 x .95) b. (.9 x .5) + (.1 x .5) c. (.9 x .95) +(.1 x .95)

10. Continuous gene flow between them will eventually result in the following frequencies:
    - a. Pop. A: p = .1, Pop. B: p = .9
    - b. Pop. A: p = .4, Pop. B: p = .6
    - c. Pop. A: p = .9, Pop. B: p = .1
    - d. Pop. A: p = .5, Pop. B: p = .5

Answers to Quizzes

Quiz 1

1. c.

The term "gene" most commonly refers to a section of the DNA molecule which codes for a specific protein; hence, a sequence of nucleotide pairs. Physically, this is the "locus" on a pair of chromosomes. Often, "gene" is used as "allele", but this is inappropriate.

2. a.

3. c.

4. d.

Paired chromosomes occur in all normal body (somatic) cells and in zygotes, the fertilized egg. In gametes (sperm and egg) only one chromosome from each pair is present.

5. a.

Since gene loci are found at specific points on chromosomes, where chromosomes are paired so are the loci.

6. d.

The paired locus contains two "alleles" -- the individual units of genetic expression. Different expressions of a gene (Type A v. Type B blood at the ABO-locus, for example) result from the presence of different alleles.

7. b.

The genotype is the paired genetic information at the locus -- the two alleles. Sometimes "genotype" is used to define the sum of all genetic information for the individual or cell, more appropriately called the "genome".

8. a.

The phenotype is the manifested characteristic determined by the genotype, or paired alleles. Sometimes it is used to define the sum of all manifested traits for the individual, more appropriately called the "phenome".

9. c.

A genotype with both alleles alike is homozygous (homo = "alike").

10. a.

The allele which, in expressing itself, prevents the other allele's expression, "dominates" that expression, and hence is dominant.

11. b.

"Hetero" = "different".

12. d.

The only way a recessive allele may be manifested in the phenotype is if it is paired with another of its kind -- in a homozygous genotype.

13. c.

If two different alleles at a locus are each manifested in the phenotype, they must each be equally "dominant", or "co-dominant".

14. c.

Meiosis results in mature reproductive cells (gametes), each containing one-half the normal chromosome number (1n, as opposed to 2n -- the paired chromosomes).

15. a.

A heterozygous locus contains a pair (2n) of unlike alleles; meiosis results in gametes (1n), each with one of these different alleles.

16. b.

During meiosis, the paired alleles at every locus are segregated into different gametes.

17. a.

If the two loci are close together on a single chromosome pair, the two loci will segregate together, not independently, resulting in only two different kinds of gametes.

18. c.

If the two loci are on separate chromosome pairs, the alleles at each locus segregate into different gametes, and the two loci assort independently of each other, resulting in four different kinds of gametes.

19. d.

Regardless of the number of different alleles available for a characteristic, no individual has more than a pair of alleles for the characteristic.

20. c.

One heterozygous locus produces 2 different alleles, and two different kinds of gametes; two heterozygous loci, each producing 2 gamete types, produce a total of 2^2 different gametes; three heterozygous loci, each producing 2 gamete types, produce a total of 2^3 different gametes. Hence, **n** heterozygous loci, each producing 2 gamete types, produce a total of 2^n different gametes.

Quiz 2

1. a. 100 (AA) b. 100 (OO) c. 250 (AA, BB, OO) d. 200 (AA, AO) e. 100 (AO)

2. a. .25 (100/400) b. .25 (100/400) c. .50 (200/400)

3. d.

4. c.

5. a.

6. d.

7. a. .16 (160/1000) b. .26 [(160 + ½(200))/1000] c. .74 [(640 + ½(200))/1000]

8. a. .74 (TT + ½Tt) b. .26 (tt + ½Tt) c. .80 d. .40 e. No

9.
   a. .60 (AA + ½AO); Yes, because AA also = .60
   b. .50 (tt + ½Tt); Yes, because tt also = .50
   c. .26 (rr + ½Rr); No, because rr .26
   d. .70 (MM + ½MN); Yes, because MM also = .70
   e. .58 (PP + ½PQ); No, because rr .58

10. a. .25 ($.5^2$) b. .48 [2(T x t)] c. .49 (R^2) d. .18 [2(A x a)] e. .64 (O^2)

11. a. .5 (2pq) b. .16 (t^2) c. .09 (r^2) d. .19 [A^2 + 2(Aa)] e. .36 [B^2 + 2(BO)]

Quiz 3

1. d.

Selection operates through differences in either mortality or fertility. Such differences cause changes in the frequencies of phenotypes, and alleles, in subsequent generations. Simply a higher frequency (c.) indicates nothing, since that frequency may be stabilized, as in a Hardy-Weinberg population.

2. b.

The phenotype with the higher mortality rate is under negative selection. Its frequency will decline each generation as a result.

3. a.

The two genotypes contributing to one phenotype must be the heterozygote and one of the homozygotes. The one allele common to both genotypes must be dominant. The other phenotype, by necessity, consists of the remaining homozygous genotype -- the recessive.

4. a.

The phenotype with the higher fertility rate is under positive selection. Its frequency will increase each generation as a result. Relative to the other phenotype, its fitness value = 1.0 .

5. a.

The single genotype producing the phenotype with highest fertility is the recessive homozygote. This recessive trait (phenotype) has the highest fitness, so the other phenotype must include both the heterozygote and the other homozygote: the dominant phenotype. Selection is operating against this phenotype.

6. a.

The fitness value is the opposite of the selection coefficient: a phenotype with a selection coefficient of .4, for example, has a fitness value of .6 .

7. b.

Selection against a dominant phenotype removes the dominant allele in each genotype in which it occurs, eventually eliminating it entirely. Hence, the other allele, the recessive, eventually becomes "fixed" at 100%.

8. d.

Selection against a recessive phenotype systematically removes only the recessive homozuygote. The recessive allele in the heterozygote is not affected: the dominant allele protects it. Therefore, selection against a recessive trait can never completely eliminate the recessive allele.

9. b.

The rate of allele change is related to the selection intensity: the higher the value, the greater the rate of change. If the mortality of individuals with the trait increases, the selection coefficient is higher. As the number of generations increases (with the same selection coefficients), the rate of allele change actually slows down in most cases. Run Exercise 4 again and see for yourself.

10. d.

Selection operates against p^2 which, since it is the only phenotype with reduced fitness, is most logically the recessive homozygote. Selection against a recessive cannot eliminate the recessive allele.

11. a.

Selection operates against p^2 and 2pq. Since these together constitute a single phenotype, it is logically dominant. Selection against a dominant trait will eliminate the dominant allele. Thus, the recessive allele will become fixed at 100%.

12. d.

This is a "balanced polymorphism": selection against both homozygotes gives the heterozygote the reproductive advantage. Since the heterozygote has both alleles, neither allele can be eliminated. However, since each allele in its homozygous genotype is at a disadvantage, an equilibrium is achieved at which the alleles and their three genotypes remain stable.

13. c.

When the heterozygote is removed (through mortality or reproductive failure) each of the two alleles is removed. Since the two alleles begin, before selection, at equal frequencies in this population (.5 and .5), selection removes them in equal proportions. Therefore, no frequency changes occur. If the allele frequencies were not equal, however, the allele with the lower frequency would decrease. If this is unclear, try some examples using Exercise 4.

14a. c.

The first two types reduces variation by either eliminating one allele or consisteltly reducing it frequency to near zero. The last type maintains variation only if the allele frequencies are each 50%. The third type is a balanced polymorphism: it consistently maintains both alleles in equilibrium, thus insuring continued variation.

14b. a.

Selection against a dominant trait will eliminate the dominant allele, and thus the variation at this locus.

Quiz 4

1. d

Any contribution of genes to a population from outside the gene pool constitutes gene flow.

2. b

The introduction of genes from one population to another makes the two populations more genetically similar.

3. c
4. c
5. b
6. d

These four questions regarding a hybrid population all illustrate the same principle: the proportionate contributions (weighting factors) are equal (50%-50%) when the two merging populations have the same size gene pools. The resulting frequency is the average of the two:

(Pop. A x .5) + (Pop. B x .5) = Hybrid Pop. C

7. b

For 10 intergroup marriages, half the couples reside in A and half in B. Thus 5 of 10 marriages bring 5 persons from B to A; and 5 of 10 also send 5 persons from A to B. An intermarriage rate of 10% is therefore a gene flow rate of 5%. (Logically, if A and B exchange through intermarriage, for every marriage where the couple resides in A there muyst be a marriage where the couple resides in B. Otherwise, one population would increase its numbers at the expense of the other.)

8. c

With 5% gene flow into Population A, its next generation will have a gene pool composed of 95% of its original genes and 5% genes from Population B.

9. c

10. d

Continuous gene flow between these populations will have the same result as if they merged into a single population. Running the exercise will show how long this will take.

IMPORTANT: PLEASE READ BEFORE OPENING THIS PACKAGE
THIS PACKAGE IS NOT RETURNABLE IF SEAL IS BROKEN.

West Services, Inc.
610 Opperman Drive
Eagan, MN 55123

Understanding Human Evolution
LIMITED USE LICENSE

Read the following terms and conditions carefully before opening this diskette package. Opening the diskette package indicates your agreement to the license terms. If you do not agree, promptly return this package unopened to West Services for a full refund.

By accepting this license, you have the right to use this Software and the accompanying documentation, but you do not become the owner of these materials.

This copy of the Software is licensed to you for use only under the following conditions:

1. PERMITTED USES
You are granted a non-exclusive limited license to use the Software under the terms and conditions stated in this license. You may:

a. Use the Software on a single computer.
b. Make a single copy of the Software in machine-readable form solely for backup purposes in support of your use of the Software on a single machine. You must reproduce and include the copyright notice on any copy you make.
c. Transfer this copy of the Software and the license to another user if the other user agrees to accept the terms and conditions of this license. If you transfer this copy of the Software, you must also transfer or destroy the backup copy you made. Transfer of this copy of the Software, and the license automatically terminates this license as to you.

2. PROHIBITED USES
You may not use, copy, modify, distribute or transfer the Software or any copy, in whole or in part, except as expressly permitted in this license.

3. TERM
This license is effective when you open the diskette package and remains in effect until terminated. You may terminate this license at any time by ceasing all use of the Software and destroying this copy and any copy you have made. It will also terminate automatically if you fail to comply with the terms of this license. Upon termination, you agree to cease all use of the Software and destroy all copies.

4. DISCLAIMER OF WARRANTY
Except as stated herein, the Software is licensed "as is" without warranty of any kind, express or implied, including warranties of merchantability or fitness for a particular purpose. You assume the entire risk as to the quality and performance of the Software. You are responsible for the selection of the Software to achieve your intended results and for the installation, use and results obtained from it. West Publishing and West Services do not warrant the performance of nor results that may be obtained with the Software. West Services does warrant that the diskette(s) upon which the Software is provided will be free from defects in materials and workmanship under normal use for a period of 30 days from the date of delivery to you as evidenced by a receipt.

Some states do not allow the exclusion of implied warranties so the above exclusion may not apply to you. This warranty gives you specific legal rights. You may also have other rights which vary from state to state.

5. LIMITATION OF LIABILITY
Your exclusive remedy for breach by West Services of its limited warranty shall be replacement of any defective diskette upon its return to West at the above address, together with a copy of the receipt, within the warranty period. If West Services is unable to provide you with a replacement diskette which is free of defects in material and workmanship, you may terminate this license by returning the Software, and the license fee paid hereunder will be refunded to you. In no event will West be liable for any lost profits or other damages including direct, indirect, incidental, special, consequential or any other type of damages arising out of the use or inability to use the Software even if West Services has been advised of the possibility of such damages.

6. GOVERNING LAW
This agreement will be governed by the laws of the State of Minnesota.

You acknowledge that you have read this license and agree to its terms and conditions. You also agree that this license is the entire and exclusive agreement between you and West and supersedes any prior understanding or agreement, oral or written, relating to the subject matter of this agreement.

West Services, Inc.